D0242068

Underlands

Also by Ted Nield

Supercontinent: 10 Billion Years in the Life of Our Planet

*Incoming!: Or, Why We Should Stop Worrying
and Learn to Love the Meteorite*

Underlands

A Journey Through Britain's Lost Landscape

Ted Nield

GRANTA

Granta Publications, 12 Addison Avenue, London W11 4QR
First published in Great Britain by Granta Books, 2014

A CIP catalogue record for this book
is available from the British Library.

1 3 5 7 9 10 8 6 4 2

ISBN 978 1 84708 671 6

Typeset by M Rules
Printed and bound by CPI Group (UK) Ltd, Croydon, CR0 4YY

For all my forebears; but especially
Megan Nield (née Bowen) 1920–2007
and
Edward William Nield 1920–2013

Each blade of grass has its spot on earth whence it draws its life, its strength; and so is man rooted to the land from which he draws his faith, together with his life.

Joseph Conrad, Lord Jim

Contents

Introduction

Brownfield

... land that at some point was occupied by a permanent structure. In a brownfield project the structure would need to be demolished or renovated. Today, the term brownfield project is used in many industries to mean a project based on prior work, or to rebuild (engineer) a product from an existing one.

Webopedia, accessed 15 August 2013

Happy Valley is a public park in Llandudno, Wales. It sits close by the pier, on the steep south-east slopes of the Great Orme – a huge limestone bunion jutting into the Irish Sea, dominating the resort, which, since the mid-nineteenth century, has grown up on the low isthmus connecting it to the mainland. It is a Victorian seaside pleasure garden much like any other, with lawns, rockeries, terraces, litter bins, benches, exotic trees and flowering shrubs. But Happy Valley is something else too. It is a post-industrial landscape, a former quarry complex, where men once mined limestone from the flanks of the Orme.

For most of human history, the Great Orme – Y Gogarth to give it its Welsh rather than its Viking name – has been a centre of industry, with traditions of quarrying and copper mining dating back to the

Bronze Age. After 1848, times became hard for mining but fortunately for Llandudno, the new cult of tourism grew into its shoes. By 1887, this economic changeover being all but complete, the local magnate Lord Mostyn shrewdly disposed of his defunct quarries by making a gift of them to the town in celebration of Queen Victoria's Golden Jubilee. Happy Valley grew out of the ruins of heavy industry.

After a century and a half, few traces of Happy Valley's former use now remain. The surfeit of stone-walled walkways, bridges and flights of steps hint that its designers were faced with a superabundance of on-site building materials, cheaper to use than to cart away. But the most dramatic evidence for the area's former use is the old limestone mine gallery, which pierces a rocky bluff in a side-branch of Happy Valley – the 'Elephant Cave'. So called, I have always assumed, because its great void is supported by four massive legs of grey stone (three framing its entrances, and one inside). This triple-mouthed grotto is anything but natural – but it has this in common with almost every other corner of our densely populated islands.

When the last Ice Age ended, between 12,000 and 10,000 years ago, Britain's landscape bore the marks of erosion by glaciers and the effects of standing ice-sheets. Increasingly since that time it has come to wear the mark of human influence, mainly those indispensable activities of farming, mining and quarrying. But although these industries remain essential to our survival, in Britain something has changed radically since the end of the last World War. For while farming is still to be seen almost everywhere, quarrying and mining have almost vanished. Holes in the ground, whose presence has always reminded us where things came from and how everything needful that we cannot grow has to be dug up, are closing like sea anemones. Soil creep and vegetation are smothering the Earth's bones again. Transport is cheap, so rather than pay ourselves to dig rocks safely and sustainably close to home, it makes economic (but no other) sense to ship minerals from the other side of the globe, where people will work for a pittance in unsafe conditions and companies operate under little or no environmental regulation.

While I was engaged in writing this book about our geological heritage, my father died aged almost ninety-three. Reflecting on his life, and its impact on mine, I began to realize that our collective debt to the richness of the past is mirrored by a personal debt to our ancestors, who dedicated themselves to making life easier for those who came after them. If we wish to do our best for all our children, we must act in ways that will lead them to revere us, as we revere our forebears. We are stewards of a land we have inherited, and which we hold in trust for our descendants.

Outside Elephant Cave, the ground is gullied and disturbed. Several large quarried boulders lie about, and in 1928 my father was photographed on one of them. Aged about eight, he stands to attention on

Ted Nield Sr (left, in 1928) and Ted Nield Jr (right, in 1964)
at Elephant Cave, Happy Valley, Llandudno.

the spot to which he has been lifted by my grandfather. Behind him looms the cave, like a stage-set gateway to the underworld. This was his first holiday away from his native Manchester streets, and he wears short trousers fastened by a snake belt, clutching a stick in one hand and what looks suspiciously like one of Happy Valley's municipal roses in the other. Slack woollen socks droop round his ankles.

Thirty-seven years after this picture was taken, I was brought to that place on my first holiday away from South Wales, where my father had moved during the war. I was about the same age as he had been in 1928, and I too was placed on the rock for a picture. The boulder looks, possibly, a little shorter and its grey lichen spots are certainly somewhat larger; but everything else looks much the same, despite the passage of nearly four decades.

I went back in 1985. By that time I had a degree and a doctorate in geology, and a job in the oil business not far away in Chester. Llandudno was a day trip, and I went there hoping to find that rock and have my photo taken on it again. In the interim I had forgotten much about Happy Valley, but some instinct drew me to the right spot, along a path leading up a side-valley that was now much less open and bare than it had been in 1928. The path, visible clearly as a wide trackway in the old photo, had once been a quarry road, along which limestone blocks extracted from the gallery were trundled out. It had narrowed to a footpath, leading directly to the main maw of the mine. I knew that to find the stone, that fixed point within the stream of time, I had to enter this cave and emerge through one of the side-entrances visible in the background.

On the bare earth floor within, I found the remains of a fire within a circle of stones, and a few gummed-up polythene bags left by glue sniffers who had gone there to inhale vapours, see visions and speak in tongues – unaware of the thousands of years of oracular tradition in which they followed. I climbed out through the broad side-opening into the gulch beyond, and there I found that, since 1964, the prospect had changed dramatically. The narrow valley, which in 1928 was rough open grazing (and which had developed just a few shrubs by 1964) was now choked with saplings and thickets. The famous boulder, when I eventually found it, had shrunk almost to nothing. A rock that in 1928 had stood a good metre tall in open ground, and which was perhaps half that high when I first stood on it, had by the mid-1980s become almost buried by downslope erosion, growth of soil and turf.

Ted Nield Jr at Elephant Cave, Llandudno, 1985.

In 1985 I had wondered, as I searched among the thickets for the stone, whether I should have a son one day, and whether I too would bring him here to continue this photographic series. I also wondered whether this hypothetical son (who, while he might never stand on the boulder, might instead stand on its grave) would also be called Edward, like me, my father and his father (and even my grandfather's grandfather). Such, at least, had always been my assumption.

Today, a quarter-century further on again, I know that I shall never face either dilemma. The posed photos in Happy Valley shall end with me, buried by time like the stone itself under the bines and tussocks. One small personal tradition shall cease, as traditions do; even those that have lasted thousands of years, like the extractive industries that were once seen everywhere but which have now all but vanished, and among whose ruins we walk.

It is because Happy Valley is a brownfield site, a place put to new uses, that I feel there a desire to keep faith with the past most strongly. This instinct, which has run like a thread through so much of my personal and professional life, has much to do with my being the latest custodian of a name passed on to me by all my fathers. My life has never been entirely my own: it was handed on to me, in trust, just as land is handed down and reused by new generations.

The previous Ted Nield was an unusual man; though as a young boy lacking any point of comparison, I had no idea *how* unusual. I discovered it slowly, through experiences not unlike that recounted by Charles Darwin's son Francis, when one of the great scientist's offspring demanded, on visiting the home of a friend: 'But where does your daddy *do his barnacles*?' I gradually learnt that other boys' fathers did not spend long hours designing transmitting devices and speaking to people on the other side of the world at odd hours of the night, nor building their own television sets, dismantling internal combustion engines, teaching themselves to play musical instruments, learning Greek for fun or building cabinets with dovetail joints so perfect you can run your finger across them and feel nothing.

Last in this long line of Ted Nields, my life is not a piece of virgin territory on which I may build anything I like. It is more like a redevelopment project, adapting and repurposing an existing structure entrusted to my care – rather as my father's perfectly preserved 1930s toys were, toys with which I played but over which I never felt rights of personal ownership. I was not the only one to feel this way. One day, during his last years, my father explained to me that he had never regarded me as a separate person in my own right, distinct from himself, like other people. He was not being cruel – he thought I would be flattered. And he only made the observation to explain why our occasional disagreements put him into such turmoil. Any discord between us did not compute. It is illogical to argue with the past. It has always been my feeling that, however much we may like to think otherwise, the past owns us almost as entirely as the rocks beneath own the landscape. This, I realized, was a personal, as well as a geological fact.

Human beings have always turned the landscape to different uses through time, some conforming to, even dictated by, the deep grain of rocks beneath; others running contrary to geology's faults and inclinations. But that which is written in the grain cannot, in the end, be denied; and any attempt to do so will end in failure – or disaster, as so many of my contemporaries found out when their young lives were sacrificed for the sake of such failure in Aberfan, the UK's deadliest mining accident.

With the banishment of holes in the ground, which remain the source of nearly all energy and raw materials, to places far beyond our immediate horizons, we are forgetting how much we owe to past times and vanished life. All the answers to our questions about the future of our world lie in the past, in the rocks. We must get under the skin of the land.

1

Formation

Formation. 1450. [ad. L. *formationem*; see –ATION.] 1. A putting or coming into form; creation, production ... 5. *Geol.* 'Any assemblage of rocks, which have some character in common, whether of origin, age, or composition' (Lyell) 1815.

Shorter Oxford English Dictionary

If stationary men would pay some attention to the districts on which they reside, and would publish their thoughts respecting the objects which surround them, from such materials might be drawn the most complete county-histories.

Gilbert White, 1788

In 1957, my mother and father moved into their first and last house. I was one year old. It had been built about thirty years before, around the same time as my parents were born, and it stood on a south-facing scarp of sandstone overlooking the Bristol Channel. A staircase of ridges rears up behind my hometown, hemming it in against the sea and rising to the best part of six hundred feet. These southerly ramparts of the mighty South Wales Coalfield, breasting the moist

Atlantic air that wallops up the Bristol Channel day after day, help
make Swansea Britain's soggiest city.

'Dalton' did not look as though it grew out of its native soil,
though it did. It was a fairly typical 1920s soon-to-be-suburban semi,
with a long, tapering back garden in which I first encountered the natural world. On the fringes of the countryside, the house faced a road
so new that a few of its older neighbours had been built back-to-front,
expecting a different road plan: their cottagey scullery doors greeted
the visitor and the postman, while panelled hallways and grand front
doors gave on to back gardens. It was the first road among many that,
through the 1930s, grew over fields into a maze of made-up highways – thoroughfares that, having no other history, went nowhere
except to the new buildings that lined them.

Not all the local roads were so newborn. Only one generation earlier, this area had been a meeting place among fields and market
gardens between two deep lanes, close to some farm buildings
clustered about a spring and dating to well before the first nineteenth-
century maps. Not a settlement, barely even a hamlet, these early
maps disagreed about what to call the spot, which then lay outside the
Borough of Swansea, a city since 1969, of which it is now a suburb.
The sandstone ridge, known (though never to map-makers) as Banc
Mawr, then bore upon its back a number of dairy farms. The nearest
of these was Llwyn Mawr, and its long whitewashed barn roofed in
clay pantiles gave its name to one of the two lanes running north over
the Banc. As a boy I would walk there to buy eggs through a tunnel
of hazels, hawthorns and sycamores that met overhead at a ragged
seam of sky.

The other lane ran west along the foot of the Banc, and shared its
name with a large Victorian Gothic country mansion, whose stable
wing backed on to it – Hendrefoilan House, built in 1855 by industrialist, mayor and local MP Lewis Llewelyn Dillwyn (1814–92). An

amateur geologist like his father, Lewis had married Elizabeth ('Bessie') De la Beche, daughter of Sir Henry Thomas De la Beche (1796–1855), first director of the British Geological Survey. (Indeed, so much did her adopted family admire the great geologist, they even saddled their own son with the tongue-mangling moniker Henry De la Beche Dillwyn.)

The meeting of these two lanes came to be lumped together with the village of Tycoch, though in the late nineteenth century the now vanished Tycoch Farm lay a quarter-mile to the east. This road junction, near one of the many springs that rise in a line along the Hendrefoilan Road and dot the hillside, was given the puzzling sobriquet of Garn Las (now Carnglas). The name means 'Blue Tumulus'. Goodness knows where this alleged tumulus was, or how it came to be blue, but few if any local names find any physical echo in the pleasant but rather faceless suburb that Tycoch has become in the half-century and more that I have known it, and during which a great and unquiet revolution has overtaken it and everywhere else.

As I grew up and saw things changing for what, to me, was the first time, I began to wonder about how they had been *before* I appeared on the scene. In 1957, the tree tunnel of Llwyn Mawr Road reached almost into the heart of the growing village, where a few Victorian houses and an Edwardian school had since been built. Our street, named with an absurd grandiosity that I failed to appreciate until I saw the real Bayswater Road, broke through its high, wooded banks on the west side. Much wider than the lane it joined, the upstart street had made the town planners' intentions clear: the days of immemorial lanes were numbered. But for the time being, on the corner there still remained one wildly overgrown vacant plot, its ancient may-hedge boundaries butting on to the privet that picked out new gardens like ours, parcelling up what had once been an orchard and market garden east of a smallholding called Ty'r Seiri ('House of Craftsmen').

Memory, preserved by names and traditions, roads, ways and boundaries, outlasts material things, and because this long-vanished dwelling was last inhabited by a woman called Nettie, the brambly copse that had evolved among the ruins of her house was always known as 'Nettie's'. Nettie's overhung the western limit of my new outdoors. Her cottage had sat in the hedged-off corner of a larger field, which had been her smallholding. The lower reaches of our back garden (which then seemed to me to reach down almost to the sea, where the foghorn of Mumbles Head lowed its falling, slurred minor third) ran along this line of hawthorn, box and dog rose. As the garden fell away, the plot narrowed and disappeared among overarching trees that shaded some neglected gooseberry bushes and an ancient, cankered apple tree. Neighbouring gardens also boasted some of these surviving remnants of Nettie's market garden, faithfully indicated on the first 1884 Ordnance Survey map with shaky diagonal hachuring.

The eastern limit of Nettie's orchard had become the far hedge of the garden next door but one to ours (I ranged widely through our neighbours' gardens in those days). Long before I ever saw an old map, this fact impressed itself upon me because, distinct from the upstart privet in between, this was a proper field boundary, centuries old. Its earthen bank was built up around a core of rubble masonry and bore not only the overgrown remnants of hazel and hawthorn, but two ancient trees – I think they were elms – and a younger sycamore.

Archaeologists tell us that the most persistent, imperishable man-made objects, ranking alongside other near-indestructible things such as flints, are not physical at all; they are *boundaries*. A frequently cited example is the perimeter of the Roman villa estate of Ditchley in Oxfordshire, which it is possible – with a little education of the eye – to pick out on modern maps. This boundary, thrown up in the first century AD, still survives in the twenty-first-century landscape and for much of its length is followed by parish boundaries that themselves date from Saxon times. Whole empires can come and go, but farming continues, year on year. Farms need owners, and owners need fences.

Next-door-but-one's far hedge had a height and structure of a wholly different order from the others. It not only boasted ancient trees, but a rich diversity of species, one of the surest indications of the age of a hedgerow – a fact that first broke upon the public mind in 1965, when a biologist called Max Hooper published his ideas on the relationship between diversity of species and hedgerow age. Hooper's Law, as it came to be known (and later somewhat discredited by more detailed work), suggested that the age of a hedgerow could be estimated by multiplying the number of species of trees and shrubs by 100. At a time when concern over the indiscriminate use of pesticides and the grubbing-up of hedgerows was beginning to make itself felt (Hooper estimated that 10,000 miles were then being lost every year), it proved a powerful political weapon for conservationists – though to be fair to Dr Hooper, he never claimed the status of a scientific law for his hypothesis.

Among the trees' roots I counted bluebells and primroses and many other species I recognized but could not then name, as well as the exit tunnels of the tawny mining bee, *Andrena fulva*, common in mown banks and especially those around orchards. *A. fulva* likes nothing better than fruit blossom, and tends to fly in early summer when pear, apple and cherry are in flower. I liked to think that those bees were the lineal descendants of bees Nettie might have seen in her garden, fumbling the blooms as the great events of history about which we learn in school rolled on, far away, like distant thunder.

Everything about that hedgerow spoke of an inherent pastness that, I began to feel, was maybe going on, right now, in this place –

though invisible, beyond the veil. I remember for the first time sensing a kind of temporal vertigo, brought on by seeing in my mind's eye, simultaneously, pictures of a place in all the different guises it has worn through the whole of history, staging posts along the endless but continuous river of time, which flows not through the two dimensions of a map, or the three of a real landscape, but through all four dimensions: length, breadth, depth and – most importantly – duration. The present is just one toe-dip in this endless flow, which is running still in the high hills of the distant past and flowing unbroken to us in the here and now. The human mind, I realized, was the only sense organ that has ever evolved the ability to see through time.

According to a map published in 1900, Nettie's cottage was still standing, while across Llwyn Mawr Road was another ancient small-holding, aptly called Dan-y-Rhiw ('Foot of the Hill'), wedged in the crook of the junction with Hendrefoilan (which has nowadays been more correctly respelt 'Hendrefoelan'). I more than remember this ivied ruin, its gable end lost in creeper and haunted by rooks – I constructed it. Impossible as it was to see beyond the wall, I had to supply from imagination what might lie there.

I built, in my mind's eye, a concealed ruin of romantic, even epic proportions, worthy of Caspar David Friedrich or Sir Walter Scott: '... the howlit crying out of the ivy tod ... the moonlight coming through the auld windows'. I can picture it still, and even now when I read gothic passages from novels such as *The Antiquary*, it is to this imagined place that my mind returns. Filling gaps in the evidence with imaginings, an act that comes as second nature to children, usually withers away in adulthood but lingers among certain scientists, especially the sort who must deal with facts that are always fragmentary – the scientists of the past – I now see that my non-existent, hypothesized ruin was a forerunner of the pictures I would build in a more constrained way as a geologist, trying to stretch as far as possible the sketchy evidence that time has passed down to us, without breaking into unbridled fantasy.

According to the late Mr Jack Williams, who was born there during the cottage's latter days, his own parents (eager perhaps to join in the area's growing urbanity) had built an unremarkable bay-windowed villa in their back garden and allowed their ancestral home to decay into the state I remember. Its remaining gable end finally toppled one night, during a powerful westerly gale that dragged at the leaves of its ivy crown and pulled it over, revealing the poverty of what really lay beyond it. Soon after, its rubble fell victim to municipal improvements of the late 1960s, when planners installed concrete lamp posts and cheap breeze-block retaining walls, effacing the past and robbing 'Blue Tumulus' of its last claim to rural character.

All this lay far in the future when the Ordnance Survey came around again in 1938. Dalton had stood for a decade, gleaming in limewashed roughcast next to Nettie's corner-plot wilderness, staring fixedly into the new street with which it conspired in trying to ignore its disreputable neighbour. Yet beyond our front gates, reeking of creosote in the sun, and over the whitewashed garden wall that rimmed the lawn like a stiff collar, Dalton's suburban pretensions were swiftly punctured. By day, up and down the ragged lane, carrying mysterious stone flagons, farm labourers in slack greasy braces, their collarless shirts open to expose yellow chest-hair and beetroot skin, would slouch, cough and spit. They never seemed to look into our road, as though it were invisible to them. I used to think that they were people of another time, which is what they greatly resembled; I was seeing them as I sometimes imagined Nettie herself, staring at the fruit blossom from her back window.

Close to where the lanes met, a pump had once stood (marked 'W.T.' on old maps, for 'public water tap'). My mother remembered it, though it was long gone by the time we moved there. Modern maps rarely show them; but on all these early surveys two words, 'Spring' and 'Rises', dot the hillside. Well into the 1970s, many of these were still bubbling up subversively from beneath tidy front walls, dashing across pavements like escaping convicts by the shortest route to a drain. They murmured a truth – that while geography may be tamed, geology will always reassert itself. Rocks are the underground resistance; the recalcitrant grain that runs through the land beneath the wood. As the past is supposed to do, it haunts. It explains, and will not be suppressed. And those who would lie to us if they could, mistrust it.

Geology can be invoked to expound most root causes in history, and those who are elected to represent the interests of others come to learn this quite quickly. In April 2000, I attended the launch of the Parliamentary All-Party Group on Earth Sciences, an initiative of Rhondda MP and geologist Allan Rogers, aimed at providing Members of both Houses of Parliament with a forum in which earth-science issues relevant to their constituents' interests could be discussed. Under Pugin's Gothic splendour in a House of Commons committee room, the Group's inaugural President, the Rt. Hon. Gillian Shepherd MP, said: 'I may not know much about geology, but as an MP, *I know it means trouble*.' This remark surprised many scientists present, who looked comical trying to suppress their long faces. But she was so right.

Geologists often talk about 'faults in the basement': ancient fracture lines running through rocks that were already very old when they became covered by others. These basement faults, deep-seated structural weaknesses, govern what happens above them in later Earth movements and so lend the land an immemorial grain. Families have faults in the basement, too. I see that my boyish enthusiasm for holes in the ground, and my tendency always to see the present in terms of what went before, and perhaps even my impatience with attempts to make me do what I didn't want (by the tidy-minded town planners of education) are immemorial in the same way. Indeed, I suspect that my mining, quarrying, stone-dressing journeyman ancestors – of whom there turn out to have been an inordinately large number – had them too. Everything is older than we think.

Subversive springs follow the foot of the sandstone ridge of Banc Mawr like a necklace, one having been pressed into service in 1855 as the water supply for Lewis Llewelyn Dillwyn's new mansion. I remember finding that spring as a boy, still bubbling into an open brick cistern upslope from Hendrefoilan Road where it passed behind the stable wing of the big house. Rainwater, percolating down through the Pennant Sandstone, was meeting an impermeable layer and working its way to the surface all along the foot of the hill. It was the first spring I had ever encountered. I plunged my hand in, enjoying its natural rebellion.

With their geological backgrounds, Lewis and Bessie Dillwyn would have known that these springs beneath the sandstones of the Pennant escarpments were there because they were underlain by impermeable shales, often accompanied by coals. And so, while stone is quarried from the Pennant Hills, coal is won and iron-rich shale dug for bricks in the valleys between. These shales, weathered for a year, mixed with water, moulded and fired with the local coal, typically made heavy bricks of a deep red colour described by building historian Alec Clifton-Taylor as having 'an almost unbearable density'.

Today, Hendrefoilan House is owned by the University of Swansea, the stable wing housing the South Wales Miners' Library. The house itself, hemmed in by overgrown trees planted by Dillwyn himself ('Planting at Hendrefoilan' being one of his laconic diary's most common entries), houses the Department of Adult Education and is faring less well. Its walls are dank and green; unfortunate outbuildings have been added; two stainless-steel flues mar the facade. Behind grubby windows, stacked chairs give the impression of an abandoned asylum.

The trees now block the view from the house, which in 1855 would have been one of uninterrupted countryside stretching down as far as the broad sweep of Swansea Bay and the twin islands of Mumbles Head. This is perhaps not to be regretted, for they screen off housing estates and the University's student village. Standing on the lawn today where Bessie Dillwyn stood in 1858 to sketch a view of the house and Banc Mawr rearing up behind it, one finds that trees block that vista also. Then, as now, the hill was not being actively farmed (Dillwyn probably left them fallow to protect his water supply) and the fields have slowly returned to nature. However, in Bessie's sketch one can clearly make out the back face of a quarry just to the right of the house – from which much of its stone was extracted. This was the first quarry I ever entered.

Dillwyn may have been a wealthy man, but his roots were Quaker. For a member of the new industrial gentry he lived a notoriously frugal life, and certainly understood the folly of transporting mate-rials long distances when perfectly serviceable dimension stone underlay his own land, a little way uphill from the build. Indeed, with the exception of the Normans (who after the Conquest often sent home for their favourite Jurassic limestone, the Pierre de Caen, if it could be landed directly from boats), the long-distance transport of stone and all its attendant effort and expense was an option that only began to be available to builders in the eighteenth century, just as styles were demanding greater urbanity – finely coursed sawn stone, smooth finishes, narrow joints and sharp edges.

Before the eighteenth century, the city walls of Oxford (as well as its castle and its older churches) were largely built from local 'Coral Rag'. As its name suggests, it is not a fine stone – it is a reef limestone, packed with fossils; but it is amazingly durable. Through the thirteenth to fifteenth centuries, the demand for smooth sawn stone for use in churches, colleges and chapels was small enough to be supplied from other local quarries. But when, during the Renaissance, more ambitious buildings required entire walls to be faced with the smooth-sawn finish known to builders as 'ashlar', architects demanded a rock that had no pronounced grain and so could be cut with equal ease in all directions – a type of rock known to quarrymen as 'freestone'. Existing quarries could not satisfy this new demand, and for the first time quarries at nearby Headington were called upon to supply it. It was not up to the job and weathered very badly.

Headington Hill Hall, now part of Oxford Brookes University, was once the headquarters of Pergamon Press, owned by notorious swindler Sir Robert Maxwell – the 'Bouncing Czech' – whose many sins were compounded by his decision to publish my very first book. The stone that runs beneath Headington Hill proved no less treacherous. Oxford's medieval builders had never used Headington as dressed stone, choosing instead to lug better-quality material from quarries eight times further away. This should perhaps have given later builders pause for thought; but being Renaissance Men, they thought they knew better. And so it was that, during one of Oxford's biggest-ever building booms, college after college was extended – or even rebuilt – in Headington Stone in sawn blocks. Headington Stone is still being replaced. The academy has been paying the price of its refusal to learn from history ever since.

Although Pennant Sandstone from the South Wales Coalfield is variable in quality, it is widely used as a building stone – though almost never as ashlar, for even its most useful strata tend to possess a strong bedding lamination. Instead, it is used as coursed blocks, tops and bottoms parallel to bedding, but with their outer faces dressed roughly and generally convex, using a chisel or punch.

Masons call stones of this finish 'shoddies'. Most nineteenth-century miners' cottages were built using coursed shoddies. Dillwyn's architect went for a slightly more genteel look, ordering surfaces flat but 'punched' to a slight rustic texture.

Dillwyn's only concession to luxury in stone is seen in the dressings of the door and window openings, fine window tracery and the grand bays on the south side. These are all in Bath Stone, quarried at Box in Wiltshire, and brought to Swansea by Mr Brunel's Great Western Railway, which had arrived in town in 1850. Brunel discovered a new source of Bath Stone when excavating Box Tunnel – the first railway tunnel to be built without a brick lining – and so had extended his business interests to rock mining. Today, a hundred miles of tunnels and galleries, cut by hand until 1948, riddle the ground between Box and Corsham, from which stone was taken, not so much for use in Bath but in the decorative dressings and cornices of the better sort of buildings then shooting up all along Brunel's railway, from Swindon and Bristol in the east to South Wales in the west.

Apart from this extravagance, when Dillwyn opened his own quarry on the slopes of Banc Mawr, he was only doing on a more ambitious scale what any nineteenth-century farmer wishing to build a new pigsty would have done. He used whatever materials were to hand – either on or around the site, or quarried as near to it as possible. Nettie's house and its neighbour, in their modest way, were also assembled from materials in situ, though the stone in their rubble walls was never more than roughly dressed and certainly never sawn smooth (it would have been impossible, as only rock that is freshly hewn from a quarry is soft enough to be sawn because it retains its pore water, or 'quarry sap'). The rock-types represented in those poor farmhouse walls were much more varied.

Today, the Pennant Hills are mantled in boulder clay dating from the end of the last Ice Age. About ten thousand years ago, the hills marked the southern edge of an icy waste extending all over Northern Europe as far as the pole; and because sea level was then 120 metres lower than it is today (so much of the world's water being locked up in the ice caps), they were a good deal further from the coast.

At that time, instead of a commanding view of Swansea Bay, wolf, mammoth, hyena or human might have looked out from the top of Banc Mawr towards distant Exmoor across a vast, forested plain where the tides now run. That gloomy pinewood concealed, somewhere far to the south, the ancestral River Severn flowing towards its estuary in what is now St George's Channel. We know this because at low tide, where the thin veneer of beach sand has been eroded away, a bed of sticky blue clay crops out, gripped tight by the knuckly roots of ancient trees.

When the ice retreated, powerful rivers of meltwater cascaded for a time over the Pennant Hills, incising deep valleys that today host weedy little misfit streams, like boys in men's clothing. The hills became mantled in stiff, clinging clay filled with boulders collected by the glacier as it scraped across the coalfield. Dense, purplish sandstones from the Brecon Beacons, sugary Millstone Grit from the North Crop – all were dumped together as the torrents ran out across the plains. These exotic boulders eventually found their way into the walls of Nettie's house.

Bit by bit, my parents began to tidy their new back garden; though I soon emerged as the only keen participant in this digging-led process. And so, as well as holes for holes' sake, I dug beds for lupins and broad beans and watched as, out of every patch of rich topsoil newly upturned, Swansea's rains would wash fragments of white pipeclay, like little bleached bones. Most were of stem, but a few were still attached to broken bowls, whose heel was sometimes pointed, sometimes broad and flat. I even found a delicate mouthpiece tipped with yellow glaze.

I amassed an enormous collection, discovering in due course (from my *Collins Field Guide to Archaeology in Britain*) that clay pipes were useful for dating archaeological sites. Cheap, plentiful and fragile, clay pipes were discarded in large numbers, and from the end of the sixteenth century, their size and shape, particularly of the bowl, underwent rapid evolution. Through the eighteenth century stems became more slender while bowls grew larger, developing a chunky flat heel for resting on the hearth – which explains why so many are smoke-blackened. Through the seventeenth and eighteenth centuries, the angle between bowl and stem became less obtuse, bowls eventually standing almost upright, before the trade died in the mid-nineteenth century, not long after Mr Dillwyn began work on his new home. The pipes told me that Nettie's smallholding had been worked from at least the mid-eighteenth century. Our garden topsoil contained, jumbled together as though recording a single instant of time, a good two hundred years of pastoral life.

Here was confirmation of sorts that building habits changed little between the eighteenth century and the mid-nineteenth century. Everyone used whatever was in their backyard. However, Hendre-foilan's Bath Stone dressings afforded some hint that change was overtaking the way we sourced materials, at least for those whose wealth gave them access to powered transport.

As well as the stones of Nettie's demolished house and the pipes smoked by generations of market gardeners – perhaps even by Nettie herself – my digging also turned up bricks left over from the construction of Dalton back in the late 1920s. Bricks are much less cagey about their origins than natural rock-types, for they usually bear a maker's mark. I found two sorts. Most numerous were rough, dark red bricks with a shallow oval depression (or 'frog') containing the words 'CLYNE KILLAY'. In smaller numbers I found smarter, tile-red bricks with a deeper, rectangular frog bearing the name 'TUNNEL'.

These were respectively the house's 'commons' and 'facing' bricks – the former used in walls that would eventually be roughcast, and the latter only where they would be exposed to view – which in Dalton's case meant the chimney stack. 'Tunnel' bricks came from a works just over the Banc and a little further east, at Cockett (near Cockett Tunnel, built by Brunel to take his Great Western Railway to points further west). The commons were made at the Killay brickworks in Clyne Valley, one of those deeply incised glacial meltwater ravines carved at the end of the Ice Age in slightly older rocks from the Lower Coal Measures, rich in black shale. Both works were less than two miles away.

The mortar between the bricks was also local. Instead of the now universal Portland Cement, Dalton had been built with black lime-ash mortar made from a mucky mix of quicklime (from the nearby limestones that enclose the coalfield as a dish encases a pie) and coal ash – the waste product of the very kilns that fired the bricks. Lime burning was a huge artisanal industry, folk memory today suggesting that local farmers were still burning their own as late as the 1930s. These two native ingredients were then mixed together on site into a black slurry using a mortar-wheel, forerunner of the cement mixer. Mixed with horsehair, this unattractive but economical material was also used to plaster ceilings and walls. Neither Dalton's bricks nor its mortar had travelled further than two miles to site, and in the 1920s probably did so by horse and cart.

Even the nearby school, a building backed by municipal wealth (opened in 1908 and being demolished as I write), hardly differed. Its walls were faced in bricks from the Graig brickworks in Morriston, only slightly more distant at about four miles away. Hidden walls were built of cheaper Pennant Sandstone rubble and lime-ash mortar. Window dressings and other architectural detailing employed – just as Mr Dillwyn's house had done – the reliable freestone of Bath, a cheap train-ride away on the GWR. The school's slightly greenish slates came from the Gilfach Slate Quarry, Llangollen, in the Eastern Cleddau of Pembrokeshire, also travelling most of the way on Brunel's railway. Nothing much had changed greatly between 1855 and 1908.

You might think that a coalfield – rich in building stone, fuel, iron-stones, limestones for mortar, shales for brickmaking – might be different from mineralogically less well-endowed regions. But when it came to building, the same economic necessities applied everywhere. My north London house dates from the mid-1870s and is built from soft yellow London 'stocks'. Examine these bricks closely, and you find fragments of bone and cinders in the sandy yellow matrix. London is mostly built on clay, so brickmaking was often even more local than in my home town. It was frequently done on site, often speculatively, by itinerant brickmakers who built temporary kilns (or 'clamps') and sold to builders working nearby. Before firing, the London Clay would be mixed with sand, chalk, ash from domestic fires and bone scrap collected from back doors. Nothing saleable was ever thrown away by thrifty Victorian households. But my house's bricks probably came from brickfields in nearby Clapton – barely a mile distant.

We don't as a rule stumble across brickworks in our backyards today, so what has changed? Not the least of the revolutions that came in the wake of the Second World War was a transport revolu-tion, fuelled by cheap oil imported from the Persian Gulf. This transformed building habits that had remained unaltered for millen-nia. Very soon after my parents and I moved into Dalton, half of Nettie's remaining vacant plot was sold, and between 1960 and 1961 a new house rose on the corner. I enjoyed my 'Rosebud' time 'helping' its builders – before the dismal clouds of education descended and buried me in what other people thought was interesting and necessary instead.

The bricks for our new neighbours' house came from near Peter-borough – over two hundred miles away. The commons, functional equivalent of our Clyne bricks, were light, machine-moulded pink 'Flettons' made by the London Brick Company (and marked 'LBC' in their deep, triangular frog). Made from Jurassic Lower Oxford Clays since 1882, they became almost universal with the advent of cheap road transport. The Oxford Clay is rich in carbon, so bricks made from it partially 'fire themselves' and therefore require less fuel. With this price advantage, Flettons – with their characteristic dark 'kiss marks', where one brick touched another in the kiln – outcom-peted nearly all others. The mortar for the new house was, of course, Portland hydraulic cement, made from Jurassic limestones quarried at Aberthaw in the Vale of Glamorgan. Both bricks and mortar were trucked in on Bedford flat-loaders, burning diesel made from Persian Gulf oil. Not one element in that house's construction was sourced anywhere within fifty miles.

The last remaining plot of Nettie's wilderness remained unsold until 1977 when hawthorn, apple and pear trees were felled and burnt on a great, fragrant pyre. Nettie's ghost fled, as new footings went in and a new Ty'r Seiri rose over the old. I was studying for my geology degree finals. This time, the builders were not my friends. I watched them with a familiar mixture of hatred and envy. I envied them that they were out in the sunshine, shirts off, unburdened with revision. I hated them for destroying the wild place in which I had dug holes, and which was once such a rich little refuge for plants and wildlife. And I hated them for using anonymous concrete blocks.

These four buildings on the same acre of land – Nettie's farm, my parents' 1920s semi and its two post-war neighbours – perfectly chart our changing relationship with the rocks we use to build our world: the general removal of Earth-materials production from 'known and local' to 'remote and invisible'. Fifty years ago, everyone knew where bricks were made; and since they could see it being done might even know *how*. Now, almost nobody knows. The Clyne brick pit, still working in my lifetime, was killed off by cheap Peterborough Flettons and has returned to nature as part of an entity dubbed 'Clyne Valley Country Park', actually a post-industrial landscape dating back to medieval times. Overgrown and inaccessible, this classic geological locality (which I studied as a student) is today another rugged place in a forest, useless even as a stop on a geological field trip.

We are used to complaints about imported foods that grow perfectly well in our own country, in season – beans from Kenya, potatoes from Egypt, apples from the Cape, and so on. Earth materials are much the same. Just as in our latitudes fruit and veg only grow in due season and in the most suitable soils, Earth materials can only be mined or quarried where they occur. But they are now being moved around our planet like never before, and most of the work is fuelled by that other Earth material, petroleum. This carbon, sucked out of the atmosphere millions of years ago by living organisms and locked up safely underground ever since, is being liberated in the transportation of loads that could never have been moved when only muscle and sinew took the strain. The thrift that once dictated restraint no longer applies.

Eric Gill (1882–1940), the British Arts and Crafts sculptor, typographer and designer, described the Anglo-Saxon world as 'a handmade world . . . a slow world, a world without power, a world in which all things were made one by one . . . dependent upon human muscular power and . . . draught animals'. In many ways, that observation held true in the British countryside throughout the nineteenth century and even as late as the Second World War.

Human beings are now responsible for moving more rock around the surface of the Earth than all natural processes – wind, surf, rivers,

floods, glaciers and landslides – added together. In fact, geologists'
most recent estimates suggest that, adding the effects of increased
erosion caused by farming and deforestation, humans are currently
ten times more efficient at moving Earth materials than nature.
Anthropogenic erosion overtook natural denudation a thousand years
ago. Our erosive power has reached such a pitch that we currently
erode and transport enough rock to fill the Grand Canyon to the brim
every fifty years. Nature may have taken as long as 17 million years
to create that hole.

During a recent, gloomy pre-Christmas visit to Dalton, I took what
I hoped would be a consolatory walk to that nameless abandoned
quarry behind Hendrefoilan – the first I ever entered, and which the
daughter of the British Geological Survey's founder sketched in
watercolours in 1858. I had not been back for many years, and I
noticed immediately how much more overgrown its surroundings are
today.

When I was a boy, the abandoned fields above Hendrefoilan House
were roamed by walkers, with and without dogs. Child gangs would
play here, as I did myself – the amphitheatre of an old quarry being
the ideal setting for those games of war that feed the desperate agony
of boyhood. Today, boys have found new outlets. They either do not
wish to go outside, or are not allowed to, or both. Where I could once
have walked, I found I had to fight my way.

The quarry floor was always wooded in my lifetime; it had lain
unworked since 1855, a year before my great-grandfather, William
Bowen, was born onto these very same rocks, on the northern side of
the coalfield, and 101 years before I was born onto them, here on the
south. I remember how, from the crest of Banc Mawr, the quarry's
presence in the slope below could be guessed at only if one noticed
that those were not mere bushes rising above the bracken, but the

topmost crowns of several mature trees, their trunks rooted fifty feet below. The back face of the quarry is as sheer today as when Bessie painted it; so approaching from above, I would skirt its maw and enter via a grassed-over spoil heap extending out from the quarry mouth. No Victorian quarryman would squander energy carrying rock that would end up as waste; so stone was always worked as much as possible in the place where it was won. This short ridge of tailings represented maybe a thousand or so tonnes of offcuts, splinters and unsuitable stone too friable or full of coaly flecks to use.

Eventually I found myself in the quarry's dank hollow, under a canopy that made it too shady for undergrowth. Although less frequented, someone had been there. There was an inevitable charred log, a few beer cans, a condom. The greenish, fern-covered strata, rearing up behind, displayed their bedding – seeming at first glance to lie as horizontal as the day they were laid down 350 million years ago. In fact, this is an illusion because those beds have been folded by Earth movements, and here they were dipping directly away to the north, into the hill, plunging underground towards the coalfield.

The South Wales Coalfield is vast – stretching from Pontypool and Cwmbran in the east, through Swansea and Llanelli to its last narrow, pinched expression in the neck of South Pembrokeshire between Tenby and Broad Haven. Like other British coalfields, the South Wales field preserves its energy reserves by virtue of being a structural basin – a downfold (or 'syncline') in the strata – which, set deeper in the Earth's crust, was protected from erosion.

The coalfield is like a pie, lying in a dish made from limestones. Those limestones were laid down in tropical seas on the southern edge of an ancient island that stretched from Wales to Europe. The pie's 'filling' consists of the rocks that were deposited as this landmass, which geologists have named the 'Wales-Brabant Island', was uplifted. This was the first sign of the emergence of a great mountain belt, in whose deep roots the coalfield would hunker down safely for hundreds of millions of years. For the time being, however, this infinitely slow process (akin to the present-day collision of Africa with Europe, which will eventually squeeze out the Mediterranean Sea and raise its sediments into a new mountain belt) made itself known only by the more vigorous erosion of the rising island. Rivers draining it carried more mud, silt and sand and dumped these in what had previously been clear coral seas. A fringe of large estuaries and deltas, clothed with dense forest, began to build out from the island.

Using measurements from quarries and mines, geologists have been able to map out the thicknesses of sediment laid down at this time all over the region. These maps show how the sediments accumulating around the Wales-Brabant Island's coast became thinner away from land. Among sediments laid down near to land (which is to say, in this region, towards the north) coals make up a higher proportion of the succession because upstream areas became emergent sooner and remained so for longer. The coal forests grew longer there, and so built up thicker seams. Coals in the Swansea area are contrastingly thin because any forest that grew there lay at the deltas' outer fringes, and so was more vulnerable to drowning as the relative positions of land and sea changed.

The Carboniferous Period was the first time in Earth's history that so much of the planet's land masses became covered in jungle. Its rampant photosynthesis removed carbon dioxide from the atmosphere as plants split carbon from oxygen, built leaves and trunks from the carbon and released the oxygen as waste into the atmosphere. At that time, bacteria and fungi had not yet evolved the chemical mechanism that today enables them to break down woody plant material and return the carbon it contains quickly to the atmosphere. Atmospheric carbon-dioxide levels fell, while oxygen concentrations rose to levels much higher than they are today.

This was the reverse of what is happening now, as we burn the carbon, like coal, that was once safely locked up in rocks, so consuming atmospheric oxygen and liberating carbon-dioxide gas. Instead of the 'greenhouse' world that threatens us, Carboniferous Earth became an 'icehouse'. Ice caps developed at the poles. Glaciations came and went in high latitudes, raising and lowering the sea level in a cycle governed by wobbles in the Earth's orbit about the Sun.

As a result, the Carboniferous sea regularly rose and fell – advancing and retreating large distances over low-lying swamps, burying the accumulated peats that would one day become coal and doing so most frequently, and for the longest periods, in those areas that were then furthest from land. This oscillation generated regular sequences of rock-types in between coals as first marine, then non-marine sediments were deposited (with characteristic fossils), followed by lagoonal muds and silts and eventually river sands like the Pennant, on which a soil horizon would develop and a new forest become established. This cycle of sudden deepening and gradual shallowing was repeated, sometimes completely, sometimes partially, over and over again for hundreds of millions of years.

The marine incursions are useful to the mining geologist because they were – geologically speaking – brief events, easily picked out by the distinctive fossils that they left behind. Like those pipe fragments I used to find in the garden soil, each appearance of marine animal remains told of a single event that may have been long on a human timescale, but which to the Earth was a mere eye blink. The marine transgressions were also separated by enough time for each suite of fossils they contained to be slightly different on each occasion, thanks to evolution. And because these marine bands were created by global sea-level changes happening simultaneously the world over, they could be used to determine where any particular coal horizon sat within the geological column. Correlation of this kind, essential knowledge for the miner, was always an especially tough problem in the rocks of the coal measures, which not only varied greatly in thickness from place to place, but also derived from a complex sedimentary environment with streams, lagoons and swampy forests forever shifting about.

And so I stared at the Pennant Sandstones on that dull pre-Christmas day in the quarry at Hendrefoilan. The name 'Pennant' is Welsh; *pen* means 'head', and *nant*, 'stream'; the word evokes a place where headwaters rise. No surprise, then, that it is the Pennant Sandstone that builds the mountainous heart of the coalfield, nor that in its many springs its river tributaries rise. Most people have heard the names of at least some of these rivers, like Taff and Rhondda, because of the fame their coal-rich valleys enjoyed; and everyone who has ever visited a great British industrial city, or chuffed up the now mostly vanished railways that once ran like tapeworms through every mining valley, knows the Pennant Sandstone.

Building stones can speak loudly of certain periods in human history. The golden limestones of the Cotswolds evoke the great age of wool, from late medieval times to the rich estates of the eighteenth century. They do so because that period saw the blossoming of sheep farming in the sunny limestone belt which sits in a yellow swag across the heart of England from Dorset in the south to Yorkshire in the north. And with prosperity comes building, often grand building, and local stone becomes raised up to the praise and glory of God and of Mammon.

As the Industrial Revolution gathered pace, building moved to manufacturing centres founded on the coal basins, and Pennant Sandstone – sometimes known by stonemasons as Forest of Dean Stone, where it is of highest quality and is extensively quarried, even today – was ripped from hillsides and incorporated into miners' terraces, grim, sanctimonious chapels, virile engine houses and rail-way abutments, Victorian Gothic churches serving new urban parishes and, most grandly of all, the great nineteenth-century town halls and colleges as civic pride went mad. Jacquetta Hawkes, in her pioneering book *A Land*, wrote in 1950 that:

> There is ... something massive, enduring, grim and a little coarse-grained about these stones that seem to make them the ideal stuff for much Victorian architecture; something too about their dark greys and browns that recommends them for the town halls, exchanges, banks and prisons of our Northern towns, where native sobriety is soon deepened by a mourning veil of soot.

I turned from the quarry to the home where I had grown up, with its garden and its old hawthorn hedge, tamed now, that once marked the boundary of Nettie's house and which once probably served as her washing line (hawthorn and box having been the preferred shrubs for this forgotten purpose). There I sat for a while on the back steps, in the fleeting unreality of the present, looking out at a featureless desolation of grass. The trees and hedges, the shadows, the hedgehogs and slow-worms, the smell of fox in the far corner, the fruit trees with their blossom all felled, years ago, gone with the wasps in the windfalls.

You can see no evidence of a hole I once dug – my last, intended to be my masterpiece, deep enough to reach through the boulder-clay remnants of the Ice Age beneath the garden's topsoil to the Pennant bedrock. Alas, the going got heavier; the hole developed drainage problems and eventually, for want of a pick and a good pump, I gave up. But before that, my father had lifted his nose from the Greek text he was reading and, becoming dimly aware of heavy labour afoot, whiffled down the garden path to inquire of its purpose.

This time marked the fading of Father's two-decade-long classical period, into which he had been propelled post-war by his obsession with the writings of Anatole France, the already ruined colossus of French literature. France likewise had been captivated by classical antiquity; and while as a writer he never forgot that he was living in the present, as his biographer David Tylden-Wright put it in 1967, France's generation wished to 'carry forward, into the new France that was forming around them, the best of the old'. Now he, too, for all his Nobel Prize, has become another ruin to be discovered in the literary wilderness.

Father's classical period ended with the adoption of a fresh enthusiasm for Thomas Hardy, Dorset, medieval architecture and the Jurassic. His ancestors, though not miners, had included stonemasons, memorial sculptors and foundry men; so he too may have felt the beguiling pull of the mineral world. But unlike those physical grafters, Father always found muscular exertion puzzling.

Two wads of yellow clay flopped onto the growing spoil heap by his feet before he asked me what I had lost. 'Nothing,' I replied. 'So what are you looking for?' 'Rock,' I said. He shook his head and walked away, past the lupins and broad beans, muttering something about sending my mother along 'with blood and honey'. I wondered about that enigmatic remark until the day I too read Book Eleven of the *Odyssey*; set in the Land of the Cimmerians, shrouded in mist and cloud (and sounding not unlike a Cymric mining valley in winter) where 'never does the shining sun look down on them with his rays . . . but deadly night is outspread'.

In that land, Odysseus communes with the dead. With his sword he digs a pit a cubit in length and in breadth, and pours about it an offering of mead and sweet wine. Then he takes the sheep and slits their throats such that '. . . the dark blood flowed forth and lo the spirits of the dead that be departed gathered them from out of Erebus'. As I sat on the back steps I reflected that, almost fifty years later, my father could always remember Odysseus, but I was passing from him into the Cimmerian mist.

We have forgotten our debt to the Earth because we no longer work it or see it laid bare. We have banished mine and quarry and spoil heap, gravel pit and brickyard. We carry no coal, pump no water; we bring everything from miles over the horizon, beyond the hedges bounding our small world. We once felt a visceral connection with the Earth as we filled our scuttles with anthracite on cold winter nights, or smelt the noxious fumes of the lime burner over the fields, or saw the heat drift up from brickyards in our own backyards. That connection has been severed. Those who came before us, who braved the depths to dig out the wealth and advantages that today allow us to outsource our Earth-materials production to places out of sight and mind, lie forgotten too, their inscriptions turned earthwards.

The British countryside used to shout about our debt to the past. The holes in the ground – windows on the past that everyone used to see and know – have gone. Local quarries, brick pits, mines and gravel diggings have closed, become filled in and overgrown or flooded. Waste heaps have almost all been removed. And often, where winding gears and engine houses once stood, supermarkets with ample car parking bring us shrink-wrapped vegetables and meat, out of season, from halfway across the planet.

Valley and mountain are grown green again, and the world looks that little bit less impure. While one cannot blame a new homeowner for tidying the back garden, it is a delusion to think of this as progress. Beyond our hedgerows, nothing has changed; our debt to the past remains, and we continue to draw on its capital. In cleaning up our little patch of Earth, we have merely banished the fulfilment of our needs to places where people are too poor to afford the luxury of environmental scruples. Others who lack the means to mine and quarry responsibly pay the immediate price; but we shall pay too, ultimately. The transport of Earth materials is immensely costly in energy, and it is a price we are condemning our children to pay as an avalanche of atmospheric carbon, poised above their heads, nears tipping point.

2

Quarry (dis.)

The present is arid and full of unrest; the future is hidden from our sight. All the richness of the world, all its splendour, all its grace, is in the past.

Anatole France, *La Vie en fleur*

I sat in the Killay Brick Pit one day in my eighteenth year, watching the fireweed lose its first seeds onto the hot air that hung almost motionless in the open amphitheatre of rock. My mind was full of thoughts of the deep past. I was sitting in the hole from which the shale had come to make the bricks of the house where I had grown up, and I was there to do a project for my A-level geology exam. The pit was considered a classic locality – at least to local geologists. In those days even the grandest science seemed to grow out of local knowledge, and I had visited it on many field trips with the University's extramural department geology class, which my father and I attended on winter Monday nights.

Father's blooming enthusiasms for geology, ecclesiastical archi-tecture and Thomas Hardy had a more practical synergy, in that they drew us summer after summer to Dorset and to the fossil-rich rocks not only of its coastline, but of the scores of small and not-so-small quarries that then opened into prospects of the vanished limestone seas of the Tethys. Envisaging a future as a geologist, and that being the time when every geologist seemed to specialize in one particular period of Earth history, I felt torn. Choosing between the sparkling, golden bliss of our adopted summer home of the Jurassic and the dark drama of my native Carboniferous was not at all easy.

The rocks formed during the Carboniferous Period (355 to 290 million years ago) fall into two broad types – the limestones laid down in the balmy coral seas that prevailed during its first half, and those deposited by the river deltas and gloomy coal forests that overwhelmed them in its second. The Carboniferous always seemed to me like a doomed time – like an idyllic, pastoral eighteenth century becoming subsumed by satanic industrial catastrophe in the nineteenth. This deep history even seemed mirrored in the landscapes they produced. The lovely, Lower Carboniferous limestones of my native Wales, or Derbyshire's White Peak, generated in the land where corals lay, gave rise to beauty even under the harshest modern climates. The coal measures, on the other hand, produced nothing but forbidding moorlands and bad drainage.

By contrast, the Jurassic Period (205 to 135 million years ago) portrayed itself to me as a time of sheer earthly joy, an eternal summer paradise without a Fall, whose blissfulness communicated itself to and through the landscapes its rocks created. Jurassic marls and limestones, like those I combed for fossils at Horn Park near Beaminster, gave us Hardy's sequestered spots: small, safe valleys in a rich country of sheep, flax and cider apples, whose gingerbread church towers sentinel the gentle mists that always seem to hang there, even in July.

As a would-be geologist, I doubt that this emotional response – a sort of time–rock synaesthesia – was helpful to me. In 1975, while I and my fellow students were on field-mapping training in Cantabria, Northern Spain, *Generalísimo* Franco lay on an operating table. The Guardia Civil pointed machine guns at anything that moved, including us; an experience that tends to colour one's impressions. One evening, I suspect after too much Fundador, I confessed to one of my teachers – a very distinguished structural geologist – that I did not care much for the mountains we were studying. Unable to grasp why one set of mountains could be more likeable than another, he asked why. 'It's something about the way they think,' I said. After a pause, he asked: 'Are you sure you are doing the right subject?' But that lay in the future. As I sat in my Welsh brick pit, for all my love of Dorset I began to feel that the soil of my home represented something special to me. The Jurassic was just too beautiful. I was unworthy of it.

In the four decades that have passed since then, Killay Brick Pit and Horn Park Quarry have suffered contrasting fates. For different reasons, neither is a place you can any longer wander into and fossick about in. One is quite simply forgotten, as unfrequented as the tomb of an ancestor with no descendants, and under its covering of thicket hard even to find, let alone access. The other is imprisoned.

The rocks exposed in Killay Brick Pit were older than the rocks of Banc Mawr, on which my parents' house was built and among which my great-grandfather had hewn the 'black mystery', as coal is sometimes called, on the far side of the coalfield. These were the Lower Coal Measures, whose fine, fissile black shales were thicker than the rust-brown sandstones dividing them, betraying their origin in deeper water, further from land, at the far feather-edges of those ever advancing and retreating deltas. When they and the coals that grew on them extended into this area, they were at the limits of their expansion and so, like the Romans in Britain at the furthest edge of their Empire, held only brief dominion. The resulting coals were a mere one or two feet thick at most. Two such coals, now long mined away, had once cropped out in this very spot. One, the Farm Vein, had run under the floor of the main pit where I sat, the fossil soil on which it had grown forming the surface of a steep dip-slope that reared up behind me to the wooded rim of the pit about twenty metres above.

The coal had long been worked out, as had much of the black shale that had come after it. In front of me, in the pit's opposite face, rose a steep rock bluff exposing the edges of these shales and, standing slightly proud like the frets on a guitar fingerboard, thin rusty bands of iron-rich siltstone. On top of the shale and capping the scarp was a coarse yellow sandstone perhaps two metres thick, whose bedding planes viewed from below still showed large ripple marks created by the waters of the river that had deposited it 380 million years ago. As year by year the frost and sun did their work, and the shales frittered away beneath it, these blocky sandstones were becoming ever more deeply undercut.

A few metres up this slope of wasting shales and rusty bands was an otherwise indistinguishable horizon rich in the fossils of a non-marine bivalve called *Carbonicola pseudorobusta*, the collection of which was usually the main reason student parties visited the place. The horizon was not easy to find, despite the depredations of collectors, because any holes hacked into it quickly melted back as the shales froze each winter, soaked up the heat of the summer sun and eventually fell away to become part of the fine scree draping the face's lower slope. Such season-weathered shale had once been the raw material from which the bricks of Clyne were made, mashed up with water to a black slurry then moulded and fired.

At that time, the quarry had lain disused for two decades, and on its floor stood several piles of waste bricks, all bearing the words 'CLYNE KILLAY' or 'E & B' (Evans & Bevan, a local brewer who had for a while bizarrely diversified into brickmaking). Here and there these bricks arranged themselves into natural seats, ideal for enjoying that packed lunch which occupied most of the Belgian Army-surplus respirator bag that I used for fieldwork. Fireweed grew all over these heaps, and clumps of their feathery seed would sometimes drift into my mouth as I ate.

Those solitary days, spent noting each bed's thickness and character and searching for rare fossiliferous bands that reference books assured me were there, were among the happiest of my life. I would arrive mid-morning after the two-mile walk and stay until the sinking light grew too feeble for me to do any more. I would almost never meet another soul, or hear more than the sound of a tractor working the fields beyond the quarry rim. The cuttings and embankments that once carried the nearby Clyne Valley Railway had not yet been amenitized into a cycle route. In those days, the dismantled railway bed was a mix of ballast and fallen leaf mould that was gradually covering the ridges left by the lifted sleepers; an active deterrent to walkers. Nobody ventured far along it from the nearby remains of Killay Station, whose platforms lay, easily overlooked, at one end of a car park serving the pub – inevitably called the Railway Inn.

A mineral line, which had once taken the quarried shale out to the brickyards for weathering, led away from the floor of the quarry. Impassably muddy even in summer, it threaded under the railway embankment through a short, narrow, dripping tunnel whose entrances were faced in Pennant Sandstone shoddies. Rain or shine, it was nearly always several inches deep in water. It led via an embankment to the old brickyards, and the collieries whose coal once fuelled the kilns. This was a 'post-industrial landscape' – but only just. Clyne had fired its last bricks at the end of the 1950s, in one final spurt of hopeful activity before the post-war transport revolution made all such operations uneconomic. Nor did it help that the local mines, following their weedy coal seams, had finally ceased production decades before.

But this landscape had been 'post-industrial' for centuries, on and off. Early records reveal that the valley had been a site of mineral exploitation – chiefly coal, iron, copper and arsenic – since the fourteenth century. Only with the definitive closure of the brickworks did that part of its history finally come to an end. The area languished in what I regarded as the best of all conditions. It had lost its raw industrial edge, but was some decades away yet from having its dangers made safe.

It was a place halfway between worlds, forgotten, neglected, bitter – and potentially vengeful. Knowing that still-uncapped mineshafts and open hillside adits lay dotted about the woods excited me. The branch-choked depths of the rotting water system's many ponds, races, leats and sloughs were unfenced. No notices warned of the deep water they contained. No rusting bridges had been taped off or torn down. There was nothing tame about it. It was what it was, and you had better watch out. In that way, it seemed to have taken on something of the V-flicking 1970s themselves, with its walking-dead industries, outmoded politics of all colours, boredom and despair.

After perhaps a score of visits, my project was written. I had been relating the ancient environments in which the rocks had been deposited to the ecology of the plants and invertebrates that I found fossilized, charting the changing conditions as coal forest changed to deep water and back again, and making biological suppositions about the different species. I gave this project some pretentious title, designed to make it sound like a learned paper in the dustiest journal and pompously referring to the colour photos that I stuck into its pages as 'plates'.

Exams over, I could have gone elsewhere for amusement that summer, yet I returned to the brick pit, just to sit among the fireweed and bricks, close my eyes and imagine myself back among the channel

sands, lagoons and sandbars with their scrubby, scrambling vegetation gradually building to dense coal forests of tree ferns, club mosses
and giant horsetails – whose descendant, *Equisetum*, still grew in
boggy areas of the pit. Making pictures of vanished worlds was
always what I loved most, and this was why I was never interested very
much by such areas of geology as the mineral chemistry of the Earth's
mantle, or how the planet's core generated its magnetic field. Such
pictures had no air in them. I wanted always to consider the very
little – Blake's 'grain of sand' – and see the world shown in it; to view
the big picture from knowledge of local things; to dig a hole in a tiny
back garden's boulder-clay soil and feel the cold air of the Ice Age hit
my teeth.

I was, and still am, possessed by a 'Gilbert White' image of science
as natural history that could be descriptive, narrative, made of words
and observation and insight – like a novel, but true. I had always
hoped that in our age it might be possible to sit in a place where fate
or providence had chosen to set me down, and come to understand
the universe by extrapolation. Aged eighteen, my horizons had
widened beyond the garden hedge. Instead of digging my own holes
I was seeking out bigger ones, at least a little further afield; but the
process was the same.

The almost complete immobility of an eighteenth-century cleric
like Gilbert White, who rarely strayed or indeed looked very far
beyond his parochial patch, contrasts starkly with our post-transport-
revolution world view. But the early signs of change appeared longer
ago than 1945. Take, for example, the writings of another country
parson, Francis Kilvert, whose works first came to my attention, and
indeed to most others', in the 1970s.

Although Kilvert's diaries had been saved for posterity by falling
into the hands of the poet William Plomer in 1937, their gentle chron-
icle of the life and times of a country church mouse spanning the

years 1870–79 received their first boost towards a wider public at the hands of another champion, Sir John Betjeman. I was a little ahead of my time when it came to recidivism, so I had already discovered Kilvert before Betjeman's 1976 TV documentary put the diarist on the popular literary map. Growing Kilvertmania, driven by people's understandable desire to retreat from the despair engendered by the late 1970s into a world of sweetness and certainty, led to the release of an abridged paperback edition of his diaries.

At that point I had already read library copies of the original, Plomer-edited three-volume version, published between 1938 and 1940 by Jonathan Cape, the publishing house for which Plomer worked. And in this longer version of the diaries, I made a discovery – in passages sadly omitted from the abridgement – that, quite literally, took my breath away. Kilvert, I discovered, had actually come to within a few hundred yards of the Killay Brick Pit. It was the sort of sensation that any scholar craves. The historian A. L. Rowse described it in his autobiography *A Cornish Childhood* by quoting the Regius Professor of History at Oxford in 1942, Sir Frederick Maurice Powicke: 'It is as if you were to sit down, and find you have sat on the cat.' Kilvert's diary suddenly came alive in my hand.

In 1872, the good-hearted young curate was working in Clyro, not far from Hay-on-Wye. As well as being the high-water mark of Victorian Anglicanism, the 1870s was the decade when the first fruits of railway expansion reached as far as the Wye Valley, placing a surprisingly large repertoire of destinations within the reach of a poverty-stricken rural cleric. Even today, Kilvert's mobility seems remarkable. In addition to thinking very little of walking twenty miles to visit some ruined priory, Kilvert took many long journeys by train, thanks to the opening in 1869 of the Hereford, Hay and Brecon Railway. Along the broad and beautiful valley of the River Wye, the new railway ran six 'up' and six 'down' passenger trains every day.

And on Monday 15 April 1872 – almost exactly a century before I began to visit the Killay Brick Pit – the Rev. Francis Kilvert boarded a train from Hay North to Llechrhyd.

There is no station of that name now. Of that railway there remains only a faint trace, like a healed-over scratch through the green fields of Radnorshire and Brecknock. At the wooden filigree halt, then just a couple of years old and still smelling bright and new, Kilvert alighted and walked down onto the lower platform, which served another line, the London North Western Railway (later LMS) with its own station called – then as now – Builth Road. There, on what I came to know as the Central Wales but which has been rebranded the Heart of Wales line, Kilvert waited for an hour and a half until his next train arrived, reading a copy of *Faust* given to him by his mother. From the timetables of the period we know that he boarded at 12.45 p.m., destination Swansea.

The LNWR had been built to connect my home town with the industrial centres of North-West England, and after running picturesquely down through Mid Wales, it encountered Pontarddulais – 'the Bont' – where scenically, things began to go off rather badly. Looping then eastwards through the mining and brickmaking town of Dunvant, skirting the western edges of Lewis Llewelyn Dillwyn's Hendrefoilan estate, the line threaded through Clyne Valley to Killay and then on to the seaside, leaping the coast road on a viaduct, now long gone, that I still remember, and running into the vanished Victoria Station on tracks only metres above the highest tides of Swansea Bay.

Kilvert would have seen Clyne Valley when its industrial character was much less in doubt than at almost any other time. A farm in the valley's floor was beset with pits and their infrastructure of coke ovens, races and railways. The Killay Brick Pit would have been working, though not of course as deep as it is today, and the collieries would have been at or near their historical peak of output.

Kilvert's destination, Killay Halt, was the last-but-one stop before
Swansea Victoria, and the timetables tell us that Kilvert would have
stepped down from his train at 3.18 p.m. As smoke, steam and noise
dissipated, he saw the Rev. Sterling Browne Westhorp waiting for him
with a pretty new wagonette outside the Railway Inn. Rector of
Ilston, Gower, Westhorp had married the sister-in-law of the Vicar
of Clyro, the Rev. Richard Lister Venables, who had taken Kilvert as
curate in 1865. During Kilvert's first year there, the first Mrs
Venables had died. It was the second who provided the connection
with Ilston.

It is hard for us to grasp how deeply the Anglican clergy permeated
Victorian society, when almost every parish in the land had one
incumbent and possibly also a curate. This immense workforce provided a day job (for many, a one-day-a-week job) for a host of often
well-connected intellectuals. The Rev. Venables, for example, was able
through his brother to claim personal connections with Wordsworth,
Thackeray and Tennyson. But for Kilvert, it was his second wife's
more modest circle that brought what he most greatly valued. His
amiable character made him a welcome guest in many agreeable vic-
arages, where there seems always to have been tea on the lawn and a
wagonette for delightful excursions.

Kilvert's peregrinations took in Pembrokeshire and Cader Idris for
holidays and Brecon and Aberystwyth for church conferences. He
regularly visited Canterbury and Worthing in the east of England, as
well as the West Country; from his father's Wiltshire parish of Lang-
ley Burrell (to reach which, Kilvert records, his parishioners assumed
he would have to cross water) to Bath, Bristol, Lyme Regis, the Isle of
Wight and Cornwall. And during the ten years of his surviving
diaries, he noted the sights and smells, sounds and inconveniences of
the unheated and frequently unlit carriages of early rail transport.
There was even one excursion to Switzerland in 1869 and another to

Paris, in which city he may have met his future wife, whom he took on honeymoon to Edinburgh in August 1879. Barely five weeks after his wedding, on 23 September 1879, Kilvert died of peritonitis, and his young widow destroyed many of his diaries.

The proscribed world of the eighteenth-century country cleric had opened up immeasurably in this first flush of railway expansion. Throughout Kilvert's accounts one is struck by the sheer delight he took in railway travel. He regarded the landscape as more picturesque with a railway than without, much as I am of the opinion that there is no view that is not greatly enhanced by the addition of a noble bridge. Railways, perhaps partly because they were built in an age proud of its engineering achievements, offer the passenger a chance to observe the landscape that the motorway denies to drivers, where each sits in his own, stifling world, hurtling along through forest-screened cuttings.

Kilvert shared many of his contemporaries' opinions about the picturesque, and was conventionally revolted by the tin mining he saw around Hayle, for example: 'The bowels of the earth ripped open, turned inside out in the search for metal ore, the land defiled and cumbered with heaps and wastes of slag and rubbish, and the waters poisoned'. But by and large, Kilvert was a true Romantic: one who saw Man in the landscape and rejoiced in his works, feeling that one lent grandeur to the other. And even unregulated mining annoyed him less than the cheap tourist. But how can we reproach him for that? He saw happening to railways what we in our age have seen befall air travel – popularity.

Amid the sight and smell of industry, Kilvert set off with West-horp, the pony pulling them up across the bridge over the line, past a few cottages and up onto the Common road, crossing older and older rocks until they reached limestone country and the deep, cool valley of hospitable Ilston, in whose graveyard today, close by the east window of its thirteenth-century church (probably founded as early as the sixth century), shaded by an ancient yew, the gravestone of Sterling Browne Westhorp still stands. What neither Kilvert nor his friend could have known that afternoon was that two of Westhorp's parishioners were spending their last-but-one night with their families before entering upon a lonely eternity, deep beneath the floor of Clyne Valley in the flooded workings of Rhyd-y-Defaid Colliery.

The day of the mine accident, Wednesday 17 April 1872, saw Kilvert and friends rock-pooling after sea anemones and relaxing on the cliff amid the gorse, high above the wide sweep of Oxwich Bay, watching seagulls reeling among the rocks and over the shining waters. Kilvert, moved by the experience, recalled Genesis 3 v. 8: 'And they heard the Lord God walking in the garden in the cool of the day'; though he left unquoted the rest of that verse, which tells of how, for shame of their deed: 'Adam and his wife hid themselves from the presence of the Lord God among the trees of the garden'.

They heard about the accident when they got back to Ilston Old Rectory that evening. Water had burst in on the miners in the pit, drowning two. Kilvert wrote: 'A brave fellow had volunteered to go down the pit to look after them but with characteristic recklessness he had gone with a naked light and was blown up by firedamp and fearfully burnt though not killed. They say it will be three weeks before the water will be sufficiently pumped out of the pit to allow of the bodies being found.' In the event, they were never found.

Both victims were Ilston men, and the news travelled quickly. The accident's first mention in print came two days later. Word of mouth proved more accurate, for it was this account that was upheld at the inquest in May, held before Mr Strick the Coroner at the Railway Inn (Mr Thomas Wales, Inspector of Mines, attending). Sadly there are no records of the session beyond reports in *The Cambrian* newspaper.

We do not even know the names of all the victims – the man burnt by the explosion, the boy, the younger man, will for ever remain anonymous. Only one, William Barrett, was named in the newspaper report, and this too is an error. Coroners' inquests held in the Glamorgan Quarter Sessions archive for April 1872 report the death of William Bennett, 42, and a John Bennett, 18, from suffocation at a colliery on 16 May 1872 (also an error, as this was the date of the inquest rather than the accident). And as William had been christened at Ilston on Christmas Day 1831, he was actually only 40 years old.

Journalistic inaccuracy is of course a commonplace; but the shoddy way in which the names and circumstances of these unfortunate men's deaths is recorded, not only by *The Cambrian* but by official bodies and formal, legally constituted assizes, bears testimony to the cheapness of human life at a time when my great-grandfather had already been working underground for six years, over on the far side of the coalfield. What also strikes the modern reader are the distances people were prepared to walk.

Kilvert was happy to walk from the Railway Inn to Ilston Rectory – easily five miles, enough to give most of us good reason to call a minicab. He made that walk on a later visit, when confusion over train times had left him stranded at Killay without a lift. But the young cleric had just spent all day pent up in a railway carriage and would have been glad of the exercise. William Bennett and his son walked that same distance twice, every working day, and would have spent much of the time in-between lying on their side, hewing.

Nor were they unusual; several of Clyne Valley's seventeen known coal mines were working at this time: Killay, Dunvant, Voylart, Wern and Killan, the latter killing three men in 1924 (though then there was a proper inquiry, and the owners were brought to book for their lapses). In Clyne Valley, apart from hundreds of ancient bell pits and scores of adits, there were mines at Clyne, Ynys and the twin shafts of Rhyd-y-Defaid, first sunk in the year my great-grandfather was born, 1856, and not finally abandoned until the 1920s, shortly after he had been laid to rest in Bryntaff Cemetery, on a wet day that formed one of my mother's earliest memories. For my part, I remember the spoil heaps of these mines, tussocked with wire grass, heather and gorse, long into the 1960s.

Thinking of this part genteel, part savage history, I sat alone in the brick pit for the last time in my youth, during that reposeful summer after A levels. I remember that it was late enough for some trees' leaves to have started to turn. I had been accepted to study geology at my local university, and in this moment of transition, as one season of life changed for another, I experienced a stronger impression of temporal vertigo than ever before – and perhaps ever since. In fact, so intense was it that I feared I might be suffering some kind of a seizure, and the anxiety only served to intensify the experience.

I looked at the black shales and thin, silty sandstones rising in succession before me and culminating in the thick sandstones whose ripple-marked undersides were as fresh in the slanting light as any on a modern beach; though these were river deposits on which eventually grew a short-lived coal forest that had laid down a thin seam of organic matter that in time had become another coal like the Farm Vein on which I sat. Just as, hundreds of thousands of years earlier, the same processes had made the Lynch and the Lynch Rider, and which, hundreds of thousands of years later, would make the Curving, the New Lynch, the Yard, the Four Feet, the Big, the Amman, the Voylart and the Frog Lane and Fiery veins, exploited at the Rhyd-y-Defaid Colliery.

Eventually, those same depositional processes – maintained over millions of years – would create the Penlan Seam and the Hughes Vein, sandwiching the Pennant Sandstone on which I, my mother, her father and her father's father would be born 300 million years or so

in the future. Less than a mile from where I sat, the bones of William
and John Bennett still lay, never to be found. I fancied I heard a whis-
tle announce the pulling out of Francis Kilvert's train on its last
downgrade to the coast, over the Oystermouth Road and along the
seafront into Swansea Victoria. Reins slapped the horse's haunch as
the wagonette's iron tyres scrunched over the crushed, compacted
rock of the station drive and the unmetalled road to the 'fine, high
common' and Ilston, deep in its ravine, with its yew and its limestone
quarry, its clear, impersistent stream, its Old Rectory with its garden
gate to the churchyard and tea on the lawn.

I saw the brick pit grow deeper as the twentieth century turned; my
grandfather leaving his mining village to fight in Palestine, to be shot
by a sniper on the Nablus Road outside Jerusalem while sneaking a
night-time fag, the photo of him and his fellow convalescents in their
Alexandria hospital taken by Sister Kinrosse. I pictured his return to
his father's house where my great-grandfather, William Bowen, worn
out at sixty-six, succumbed in 1922, two years after my mother was
born in the house that he had built, below the cemetery where he was
laid. The bricks of Clyne kept selling to the speculative builders
whose whitewashed semis were attempting to bring starched collars
and civilization to new suburban streets that went nowhere but to the
houses built upon them, infilling the land leased from the estates of
the great, like Mr Dillwyn and Bessie De la Beche, their ancient acres
reticulated with privet, their garden soil full of the broken clay pipes
left by old Nettie and her labourers, whose bones lay God knows
where.

I heard the approach of trains in 1939, working hard against the
grade from Swansea on the long climb to Dunvant and the Bont,
through the hills of Central Wales, to Manchester – carrying my father,
newly appointed spectroscopy technician at the Magnesium Metal
Company, back to his parents, in compartments thick with smoke and

khaki, their saggy string luggage racks loaded with helmets and kitbags. Clyne bricks, fashioned from the shales in front of me, fired and bonded with the ash of coal won from the rock beneath, already encased the space that would, one day soon, become home to him and the girl he would take, as a blind date, to a party not far from his lodgings above Sketty, in a house on that same sandstone on which she was born, on which she would later bear me and on which she would, like her grandfather, die.

All that history, it seemed to me, was happening in the same moment, in that place. I was staring down into a seemingly bottomless pit, the parts nearest the top showing in greater detail and resolution, naturally; but its deeper, darker parts not entirely obscured far below in the echoing vaults of the Earth. Kilvert might perhaps have known an Old Testament verse that fitted this feeling, had he experienced it; I as a geologist knew the words of John Playfair, describing the time in 1788 when James Hutton of Edinburgh showed him an unconformity between two sets of rocks, the older of which had been laid and folded and eroded down before the younger were deposited over its broken ends. Realizing suddenly what a vast stretch of time this contact represented, Playfair suddenly experienced the same vertiginous feeling, writing: 'the mind seemed to grow giddy, by looking so far into the abyss of time'. I clawed my way past the bodies of the men killed at Rhyd-y-Defaid as I struggled back to the light.

It was a relief when the sensation passed, but I was unsteady on my feet and had to take my time about moving off. I reflected on my recent decision to reject more distant and ancient universities (to the incomprehension of my headmaster and no less that of my contemporaries, bent at all costs on escape). Historian A. L. Rowse described it as 'loyalty to the past', which in his case led him to keep all his old school magazines. For me, the future never beckoned convincingly. I always suspected its motives. Like Rowse (whose father was a part-time tin miner and went to quarry ironstone near Banbury during the Great War), I never liked the present very much. Rowse wrote, and it could have been me: 'It needed to become the past before it had much savour.'

So I would study in the department founded by the great Sir Arthur Edward Trueman FRS. He had died the year I was born, and was the first geologist ever to write for the general reader about the connection between geology and scenery in Britain. As a scientist, his immensely varied yet consistent theoretical work had had profound consequences for evolutionary biology, and even palaeoecology, a discipline that did not then exist. But most of all he had used the evolution of certain clams, and his knowledge of their changing ecological assemblages through time, to refine the stratigraphy of those bewilderingly complex coal measures, with their ever-changing rock-types created by the switching rivers, deltas and swamps of that critical period of Earth history.

Thanks to him, mining engineers who came after my great-grandfather's day would be able to tell more precisely their location within the sequence of Carboniferous time – and hence gain a better idea of the direction in which productive seams might lie. He did some of that work, which won him his knighthood, in this very brick pit, where he and his co-worker Miss E. Dix had found fossils typical of what they called the Communis zone of the 'Lower Ammanian'. The term is now defunct, superseded by still finer divisions made possible by greater knowledge. But here, Dix and Trueman had noted the abundant specimens of *Carbonicola communis*, *Carbonicola pseudorobusta*, *Naiadites* and even some scattered fish scales, which were in my own time still being found and collected by year after year of student geologists, brought to this hollow shrine in order to worship Professor Trueman's great achievements.

The zones he defined in South Wales were quickly found to be generally applicable over much of Britain. Idiosyncrasies of other coalfields were soon reflected in local refinements of his system, but Trueman was able later to extend it across much of Continental Europe with enormous economic benefits. As one of his successors at Swansea, Professor Thomas Neville George FRS, wrote in 1974 on the fiftieth birthday of the department: he and his research 'brought distinction to him and to the institutions in which he worked'.

How transient is fame. Just a few decades later, Trueman's photograph would be unceremoniously removed from the corridor of my alma mater and the department closed. No student boots now make the trek to the brick pit, their anticipation sharpened by foreknowledge of the great Sir Arthur. Nobody very much cares about coal, and all Trueman's work lies, if not forgotten, at least unworshipped. His memorial was dismantled.

Recently, on a freezing winter day, I retraced my steps to Clyne to see what had become of the place since my epiphany. I found that where fireweed seeds had once lifted off on the summer air from the floor of the pit, there was now a young forest of ash and hazel and pussy willow. The sandstone overhang, which before showed off its ripple marks as fresh as though they had been made yesterday, had collapsed over the banded shales below, which were no longer visible. The sequence is obscured, and the fossil-rich horizons where Trueman and Dix collected are inaccessible. So dense is the brush within the pit, one might be forgiven for not even realizing it was there. A small portal to the past is closing, the tide of time washing it flat, just as Trueman – having laid down his layer of scientific history – is vanishing from sight. Such has been the fate of men throughout time; it is now the fate for hundreds of thousands of small workings the length and breadth of Britain. In the last half-century, a pattern of working the land that had gone on unchanged for over a thousand years has slipped away unremarked.

The limestone belt of England stretches like a great sash of honour from the left shoulder of England in Yorkshire, diagonally across through the Cotswolds, to end in a complicated knot in Dorset. These are the golden Jurassic limestones of Oxford colleges and Bath crescents. They are the quintessential building stones of rural England, taking the wealth of wool from the fat pastures they underlie to produce, in the fourteenth and fifteenth centuries, the crowning glories of English parish church architecture – the Decorated and Perpendicular orders, lofty of tower, noble in conception but human in scale. Geologist Sir Arthur Trueman wrote, of the Cotswold Stone belt: 'the essential characters of this area are determined by the rocks' – a typical Trueman truism. However, the sentence he wrote next has become a lie in the years since it was written in 1938. 'These limestones,' he wrote, 'are to be seen in innumerable small quarries.'

Long before the transport revolution, people living along the limestone belt had no problem sourcing their building materials, being blessed with the very best stone in amazing quantity and diversity. Alec Clifton-Taylor, hymning the medieval stonework in this area, wrote, as late as 1962: 'What is particularly striking is its abundance ... Quarries were everywhere; almost every village had one. Most of them were quite small.' But dear old Clifton-Taylor was not one to miss, or pass over without comment, the first signs of what he judged to be a regrettable development. He went on: 'Nearly all lie abandoned and overgrown.' While that was perhaps a slight exaggeration in 1962, it is no longer.

One small but important layer in the great Jurassic limestone swathe of England is a unit known as the 'Inferior Oolite'. From the point at which it hits the South Coast of England at Burton

Bradstock, it threads north through Beaminster to Sherborne, its outcrop passing through twenty parishes in the County of Dorset. In the
late nineteenth and early twentieth centuries, 138 local quarries
exploited this unit alone, dotting its ribbon outcrop like yellow pearls.
Inferior Oolite quarries were particularly numerous for one big
reason: the unit is not thick, rarely exceeding five metres in this part
of the world, and within it the beds that are most suitable for building stone are thinner still. Yet, like the household pig, everything in
the Inferior Oolite was useful for something. Larger blocks could be
cut to smooth-faced ashlar, perfect for the local manor house or
church. Stone of lower quality would be rough-dressed for cottages,
while remaining rubble went into the walls of farm outbuildings.
Leftovers and offcuts were either crushed for spreading on the roads,
or (from the late eighteenth century) burnt to make lime for scatter-
ing on the soil.

Lime was a valuable export for all limestone regions, but even their
own soils needed liming because the cows that graze those fields
literally walk off with it as they are taken for slaughter. Cows grow
from their native soil no less than we; for every six parts of cow by
weight driven off to the slaughterhouse, one part is bone. The
calcium, carbon and phosphorus that make up that bone all come
from what the cows eat, which is grass, and the grass gets them from
the soil. In the Gower Peninsula, over 150 limekilns can still be seen
today; though of course now we have found other ways of replac-
ing these elements. All involve extensive transportation by road, and
like all modern farming, energy-hungry chemical processes fuelled
by oil. Local lime burning has vanished too, along with the local
quarry.

Agriculture is not the only use for lime. The Romans used it in
mortar, as in the encampment they built at Waddon Hill, near the
Dorset village of Beaminster. Saxon and Norman church builders

also used it. And whenever quarries in the Inferior Oolite became worked out, the quarrymen simply moved along the outcrop and opened another. The landscape bears their faint scars to this day, even visible on Google Earth satellite images if they happen to have been taken with the light at a low enough angle. On nineteenth-century maps of the Inferior Oolite outcrop, it is easy to trace quarries being opened, worked and then returned to agriculture. This went on for centuries, and when the nineteenth-century religious revival came along, the eager (and often over-eager) restorers and rebuilders of medieval and older churches were still able to call upon supplies of matching local stone. In Dorset, much of the best stone available at that time was taken from places like Barrowfield and Whetley Cross. Another was a small quarry about a mile or so west of Beaminster, called Horn Park.

In terms of quality, there is nothing at all inferior about this superbly useful rock formation, which got its name from William Smith, so-called father of English geology and author of the first geological map of Britain. He first named it 'Under Oolite', which gives a clue as to its real meaning. His pupil, the Rev. Joseph Townsend, changed Smith's 'under' to the more genteel and educated-sounding 'inferior'. But in both cases the term merely served to distinguish this lower, and therefore older, limestone from younger units above it in the succession, such as the 'Great Oolite' (Bath Stone) and the 'Superior Oolite' – a redundant term coined by Townsend for what we now call Portland Stone.

Townsend's decision to impose this Latin word upon the Anglo-Saxon soil of English has bequeathed a legacy that is still proving troublesome in our own times, two centuries later. At the end of the twentieth century, Sir Norman Foster and the Trustees of the British Museum in Bloomsbury devised a plan to return to public use its badly neglected central courtyard surrounding the Museum's famous

circular reading room. As well as closing the courtyard to the elements with a striking glass canopy, the project involved restoring the
courtyard's south portico. The rest of the building being of Portland
Stone, architectural guidelines for the material to be used in this monumental rebuild specified 'an oolitic limestone (Portland limestone
from the Base Bed or similar) to BS 5390'. The crucial words in this
contractual phrase turned out to be 'or similar'.

The successful bidder to supply stone for the project was Easton
Masonry Portland Ltd. Their tender dramatically undercut all other
bids, and with hindsight this should have been telling. The low price
had been achieved by supplying not Portland Stone but another, 'similar' Jurassic limestone called Anstrude Roche Clair. For, in our
topsy-turvy modern world, it was much cheaper for a Dorset firm to
import hundreds of tonnes of limestone from Yonne, France, than to
quarry it from under their feet. The political and public row that
ensued was the first – and to my knowledge only – major political fuss
about where stone comes from.

Anstrude Roche Clair is older than Portland limestone by about
20 million years. However, it was deposited in a very similar environment – which is why it is extremely difficult, even for an expert
geologist, to tell the two apart. The fact that rocks from different
ages but with similar depositional environments can look almost
identical is the reason why it is important to separate 'rocks' from
'time', and to follow William Smith's example by using the fossils
contained within the rocks to distinguish old from young. Without
such an independent measure of rocks' relative age, the accurate
mapping of rock-types across a landscape would be bewildering and
meaningless.

In fact, the French rock was deposited at roughly the same time as
the rocks exposed in Horn Park Quarry, which is to say during the
Middle – and not Late – Jurassic. This meant that Anstrude Roche

Clair was close enough in age for it to be described in the public prints, albeit somewhat inaccurately, as 'Inferior Oolite'. As a building stone, it is every bit as fine as Portland. Some might say it is better, with a texture that arguably appears even more urbane in a smooth ashlar block. However, when the portico fiasco was reported, media, government agencies and Parliament were caught out in a perfect storm of righteous indignation, synthetic anger and authentic ignorance, making for a rivetingly ludicrous Gilbert and Sullivan spectacle.

The Museum officially discovered the truth at a meeting in June 1999, and wisely decided to keep calm and carry on. But an anonymous tip-off to the Heritage Lottery Fund (which had made a £15.75 million grant towards the £100 million project in October 1997) blew the lid off and the cause célèbre was born. The dispute rumbled on into 2001; during the scandal a series of MPs inadvertently mis-spoke before the House, some clearly implying (having misunderstood the term 'Inferior Oolite') that this Gallic impostor was in some way substandard. *The Times* ran the loaded headline: 'British Museum restored with inferior French stone'.

The only good sense was spoken by Hugues Duflot, export project manager for the quarry owners, an Île-Saint-Denis-based company called Rocamat. He told freelance writer on all things French, Jeremy Josephs: 'Easton Masonry informed us ... that they were using our limestone on their contract at the British Museum ... its quality and technical characteristics are comparable to Portland – and I really can't see what all the fuss is about.' Under pressure to make a statement and clearly taking a different view, the Museum's managing director Suzanna Taverne said: 'We were mugged.' Chris Smith MP, Culture Secretary at the time, ordered an inquiry into whether the Museum had been defrauded, at which point management consultants PricewaterhouseCoopers were engaged.

Adding to the confusion was the new portico's fresh, unweathered appearance. It would have looked this way no matter which stone had been used to build it – Portland limestone from the Base Bed or Anstrude Roche Clair. The bright, yellowish colour of the new work had nothing whatever to do with the use of a different stone, but few people realized that, and almost none of the arts and heritage journalists who editorialized most vehemently about the affair. In fact, as the courtyard is now covered by the spectacular Rogers Partners roof, the south portico will never attain the silvery 'skull whiteness' of its older surroundings, exposed for 150 years to London weather.

Watching proceedings from Fortress House, its impeccably authentic Portland Stone bastion in Savile Row, was English Heritage, guardian of the historic patrimony. In November 2000 it rounded on the Museum, whose trustees had by then received first drafts of PWC's report and were being accused of holding up its publication. (The report was finally published in 2001, and was highly critical of the Museum's managers. However, after a subsequent Scotland Yard investigation which reported the following October, no charges were brought against any party. Nor did the Museum take any action against its suppliers.)

In its statement, English Heritage's chairman, Sir Neil Cossons, condemned the Museum, both over the stone and alleged poor workmanship, but recommended that Camden Council (the local planning authority) should not call for it to be demolished. The state-appointed heritage guardians – who, years later, did not object to the demolition of Fortress House and its replacement by a faceless glass lump – knew as little about the rocks under consideration as everyone else – stating, in an explanatory note, that 'Anstrude Roche Clair and Portland are both oolitic limestones (i.e., made out of shells)'.

A classicist would notice that the Greek root of the word oolite is *oion*, or egg; and the well-known fact that eggs have shells might explain English Heritage's confusion. But 'oolitic' does not mean 'made of shells'. Oolitic limestones may contain shells, but are so named because they are chiefly made of tiny spheres of calcium carbonate that give the rock the distinctive appearance of fish roe.

These tiny 'ooliths' are made by the snowball-like accumulation of calcium carbonate around minute nuclei in shallow, tropical seawater. Gentle currents wash them backwards and forwards across gleaming white banks – a depositional environment best described by the poet Richard Garnett (1835–1906) as the 'rolling worlds of wave and shell' in his poem 'Where Corals Lie', made famous in its setting by Sir Edward Elgar and written while the poet was working as assistant librarian at, appropriately, the British Museum.

In the Middle Jurassic, about 170 million years ago, all the land masses of the Earth were still closely associated in a single supercontinent, Pangaea, which was beginning to break up. Britain lay at about 30 degrees north, or about 22 degrees of latitude south of where it is today. Thirty degrees north was at that time well within the 'tropics' (as defined by climate) since 'the land where corals lie' then extended much further north and south of the equator than it does today. The Jurassic was a 'greenhouse' world, with no polar ice caps; and as time went on and the old supercontinent began to fragment, the rising of mid-ocean ridges on the floors of new expanding oceans such as the Atlantic caused the global ocean to spill onto continental shelves, which were covered by clear, shallow seas. Ancient, long-eroded desert landscapes that had formed slowly over 100 million years of Permo-Triassic time were slowly drowned. Archipelagos of small islets, made of older rocks and the desert sands and screes that mantled them, were gradually swamped by drapes of limestone and mud from the rolling worlds of wave and shell.

By the time the limestones exposed at Horn Park came to be deposited, these transgressions were reaching a temporary high point. In Dorset, rocks of this age are thin and full of what geologists call 'non-sequences' – breaks in the succession where long periods of time went by leaving no sediment behind to record their passing. These were minor hesitations in the great scheme of things; later, even greater marine incursions would occur, during the Late Jurassic – when Portland limestone was laid down in Dorset – and in the Cretaceous Period, when the worldwide dominance of ocean over land reached its zenith and gleaming white chalk blanketed the seabeds of the world.

At Horn Park, where the glittering Inferior Oolite sea may have been no more than a few metres deep, sedimentation was intermittent. Little or no material arrived from land. Ichthyosaurs patrolled the waters above, while on the sea floor, where ooliths slowly accreted and grew larger under the rolling shallows, clams and little sponge colonies established themselves among the countless empty shells of ammonites.

Ammonites, the extinct curly cephalopods that most closely resemble the modern pearly nautilus, were remarkable creatures. They teemed in the world's oceans, evolving with amazing speed, diversifying into an almost infinite variety of shapes and sizes, and so providing geologists with ideal time-markers for the rocks in which their remains became entombed. They are also, quite simply, the most beautiful objects. Every fossil group has its fanatical admirers; but for their sheer simplicity of form combined with complexity of detail, the ammonites seem to me to be without equal.

The rapid appearance and disappearance of different ammonites allows the Jurassic Period to be divided into very short parcels of time, called 'zones', characterized by and named after a single distinctive species. Zones are then grouped together in 'stages', and may be themselves further divided into subzones using other, even shorter-lived species than the zone fossil. The five metres of Inferior Oolite formation at Horn Park, with its non-sequences, is a so-called 'condensed' deposit. A lot of time passed here, but recorded very few syllables in the rocks. Clearly, this situation can result in great concentrations of fossils being preserved; there are cases where an ammonite 'zone' – the rocks in which a certain species occurs – may actually be thinner than the index fossil itself.

Although this extreme is not seen at Horn Park, the rocks exposed there straddle no fewer than nine full ammonite zones from three Jurassic stages: the Aalenian, the Lower and Upper Bajocian and the lowest zone of the Bathonian Stage – named after the ammonite with perhaps the wackiest moniker of all: *Zigzagiceras zigzag*. One particular bed, Number 5a (all have been lovingly logged and named by generations of geologists) is so crowded with specimens of the large ammonite *Brasilia gigantea* that it is known locally as the 'Dinner Plate Bed'. As a result of extreme condensation, this thin sliver of golden limestone, packed with the fossil remains of a vanished tropical paradise and laid bare in Horn Park, has been described as 'of both national and international importance for Aalenian–Bajocian stratigraphy, as well as a prime palaeontological site'.

Nobody knows when Horn Park Quarry was first opened, but there is no doubt that it dates from the mid to late nineteenth century and continued into the 1960s, when I first went there. Its fossils came to prominence after being described by the great nineteenth-century West Country palaeontologist (and less remarkable novelist) Sydney Savory Buckman (1860–1929) in a series of classic papers and some gigantic ammonite monographs. It is now – amazingly – the *only* permanently accessible reference section for rocks of this age in England, which gives it immense international significance for geologists. On my first visit as a schoolboy, tagging along with university extramural field trips, the site was still worked commercially, on and off. Large blocks of yellow limestone the size of family saloons lay about the quarry floor where they had been abandoned. Tall foxgloves and teasels poked up between them, their biennial life cycles hinting perhaps that things had been a little quiet lately.

Such visits were always frustratingly short – time only for a brief lecture and a quick scrabble before dashing back to the bus. To find and extract a really good specimen, an average collector would need to spend half a day, even in a place as rich as Horn Park. And so, summer after summer, Horn Park rang to the sound of hammers and chisels wielded by my father and me. A breathless quiet hung in the air, broken only as our tappings echoed in miniature the tough, physical work that had gone on there not so long before.

On such a family visit one year I obtained the best ammonite specimen I ever extracted intact – a mature example of a species first described by Buckman in 1887, *Brasilia bradfordensis*. It was twenty

The author in 1973 at work on *Brasilia bradfordensis*,
Horn Park Quarry, nr Beaminster, Dorset.

centimetres in diameter. I had spied it on an earlier visit, partially exposed near the edge of a large detached block, but my natural defeatism let me down. It needed too much time, too much skill. It was too beautiful. I was unworthy. Better that I leave her where she lay for more skilful and deserving hands. All that changed when I noticed, on our later visit, that in the meantime some idiot had stupidly walloped the exposed surface and slightly damaged the specimen. Steeled by indignation, my father and I set to work.

My long-suffering mother, like many another partner of the amateur geologist, settled down for a long wait, brewing tea behind the Zephyr. There was nothing to read, apart from regional geological guides, Volume One of the Royal Commission on Historical Monuments' report on Dorset West (essential reading for the serious churchgoer) and a copy of Florence Hardy's rose-tinted biography of her illustrious husband. Mother read that. I remember hearing her muttering, 'You dictated this to her, didn't you, you *bloody old fraud*?' Father didn't pick up on these signs, but the stock of husbands was rapidly going south.

We chipped away at the encroaching rock until the outer edge of our ammonite – very much larger now than we had imagined it to be – lay exposed. Only its adherence to the matrix below was binding the fossil to its past, but too enthusiastic a wallop in the wrong place could still shatter it and waste all the time we had invested. Finally, placing the chisel near the fossil's outer edge, one of us – I cannot remember who – struck the chisel once, twice. On the second blow, the ammonite simply popped up. I lifted it free. After almost 170 million years, *Brasilia bradfordensis* rose from the grave, resurrected entire within its own surface once again. Our cry of joy echoed around the quarry. I heard a book clap shut and a gasp of, 'Oh thank *Christ*'.

We visited Horn Park for a third time that holiday, but I found nothing special and, instead of fossil hunting, my father took the opportunity to remove the sparking plugs from the engine of his immaculate 1953-vintage Zephyr Six, then twenty years old, and check the clearances on its tappets. My mother was deputed to flick the starter on command, bringing the engine cycle to the right point for each tappet to receive its feeler gauge. Accidentally (or so she always maintained) she managed to lean on the car's column shift behind the steering wheel and engage second gear. The next time Father called upon her to flick the starter, the car's considerable tonnage lurched forward a foot or so. My father, juggling feeler gauge, ring spanner and screwdriver, dived forwards into the engine compartment, losing hold of all three. The bonnet, propped open on its telescopic strut with a lift-and-release catch, whipped back under inertia and then closed upon him like the jaws of a great ichthyosaur. Father's enthusiasm for Horn Park waned after that, and we never went back. Nor did I, until almost forty years later.

I drove out of Beaminster along the familiar Broadwindsor Road. What greeted me at Horn Park was no longer a quarry but a trading estate. Several light industrial units, faced in the local stone and not unattractive of their kind, surround a car park neatly tricked out in white parking lines and traffic arrows. The only outcrop to be seen lies in the far back corner, behind some stacked pallets and a fearsome three-metre-high palisade with spikes, placed there as though to resist the besieging army of commercial developments.

Behind that ring of galvanized steel is all that remains of Horn Park Quarry's exposures, and it is protected by more than steel fencing. It is (since 2006) the United Kingdom's smallest designated National Nature Reserve, a Site of Special Scientific Interest (SSSI, since 1977), classified as a 'Regionally Important Geological Site' and protected, for the sake of science, research and teaching, as a public amenity. Anyone wishing to collect from these closely guarded rocks

The author returns to Horn Park Quarry with *Brasilia bradfordensis* extracted almost 40 years earlier.

on the 0.32 hectares of exposure that remain must now apply for permission. Like so many public amenities – and because it is no longer actively worked, producing new exposure continuously – it is permanently locked for its own protection. Fortunately, I knew a man with a key, and he was on his way.

Horn Park's later history is instructive. When I first visited, it was owned by a Mr George Pinney of Horn Park House, and had been worked for rubble destined for new sea defences at West Bay. This was why so many freshly quarried blocks lay about for anyone to hammer in search of fossils. Thereafter, Mr Pinney's groundsman, a Mr Wells, bought the site and set up a shooting range, which I also remember from even later visits. He planned to run the quarry as a tourist attraction based on the showing and collection of fossils. Alas, the father and son he employed to find and prepare specimens were too often tempted away to the bright lights of Bridport, and the venture failed. Local geologist Bob Chandler was then brought in to help run special access for fossil hunters. Following Chandler's advice, Wells applied to the Nature Conservancy Council (as it then was) for the site to be given SSSI status.

When Wells moved away, the quarry was sold on to a Mr Gibbs, a retired fireman who had lost an eye in an accident. He sold fossils, was suspicious of collectors and charged them admission. At this time, Horn Park was still commercially worked by Gibbs's brother to supply local building projects. But eventually Gibbs too decided to sell. Horn Park passed through the hands of two large minerals companies – English China Clays and Hanson – and stood idle for some years before being sold yet again to a businessman called Philip Seal. He put the quarry under the management of his brother Ted, with plans to work it commercially.

The Seals purchased the site on the strength of consultancy advice from Southampton University on the amount of limestone reserve still in place. The surveyor, whose name I have been unable to discover but who has been described to me darkly as 'a geophysicist', seems to have had no local geological knowledge. He certainly seems never to have looked at a geological map, for he was unaware that the quarry lies on the Beaminster Fault Zone and is cut by three faults, which complicate the structural picture greatly. He advised that 'thirty feet' of limestone remained available for quarrying. To the Seals' dismay, this advice proved wildly optimistic.

It must have been an interesting day when Bob Chandler arrived at Horn Park in the company of legendary palaeontologist Dr John Hannes Callomon (1928–2010). John taught physical chemistry at

University College London, but was also one of the world's leading authorities on Jurassic ammonites. He and Bob were proudly shown a very expensive new weighbridge that had just been installed on site. The Seals had purchased this device in anticipation of the many trucks of rock that they soon expected to see trundling profitably out of Horn Park.

Chandler looked askance at Callomon, whose eyebrows were now some way over the top of his head, and broke the news as gently as he could. From his lifelong knowledge of the site and the local geology, he could say with absolute certainty that, while the quarry had once had access to perhaps a twenty-foot thickness of good limestone, nearly all of it had already been taken. Now there remained perhaps five or six feet of Inferior Oolite, but none was of good building quality. The rock below it consisted of the upper portion of a formation called the Bridport Sands. These sandstones, in a twist that later proved fortunate in court for the hapless consultant, are quite high in calcium carbonate. However, they are in no sense, or by any reasonable definition, true limestone.

The Seals sued the author of the report, calling Chandler and Callomon as expert witnesses. But the case came to nothing   once again, it appears, because of loose wording in the contract, which had merely asked for proof of 'limestone' without specifying Inferior Oolite. Leaping eagerly through this loophole, lawyers acting for the defendant cited an obscure US description of 'limestone' as being 'any rock containing more than 1 per cent calcium carbonate'. This absurd definition made 'limestones' out of the Bridport Sands, which proved enough for the judge, who dismissed the case.

Ted Seal then tried to sell the site to English Nature (a body created from the break-up of the Nature Conservancy Council, and which has risen again from the ashes under the absurd new name of 'Natural England'). Sadly nothing came of that either, so finally he decided to build the present business park, allowing Natural England to run Britain's littlest nature reserve behind the car park. The fences went up in 2004 at Natural England's expense, after a party of visiting German fossil collectors spent several days at Horn Park and, literally, filled their boots.

I was waiting at the locked gate as Sam Scriven drove up. Sam has good reason to be a happy young geologist. A Dorset man, he now works as liaison between Dorset County Council and the Paris-based UNESCO World Heritage Centre. Horn Park Quarry also forms part of his responsibilities, which include liaising with Natural England and the reserve manager, Tom Sunderland.

Once Sam had found the key, removed the heavy padlock on the bolt and worked out whether the gate had to be pushed or pulled, we found ourselves on limestone pavements which, in 2009, he and others had painstakingly prepared. These bedding planes represent the upper surfaces of beds that were turned into rock very soon after being deposited, and formed what are known as 'hardgrounds' – limestone surfaces so long exposed on a sediment-starved sea floor that calcareous cement had time to grow between the grains and lithify them. Sam pointed out how fossil calcareous sponge colonies could still be seen adhering exactly where they had grown, cemented to the hard sea floor, over 160 million years ago. After preparation and washing and the removal of vegetation, they now looked almost as fresh as they did when alive.

Sam explained that the original plan had been to erect a shelter over the entire site, into which the public could come to admire the surfaces. But the usual budgetary problems, and a few managerial ones, led to this plan being scaled back. However, the surfaces have now been beautifully prepared, removed material carefully screened for scientifically significant finds and other material collected and kept for teaching and display. Meanwhile, instead of the ambitious

Installing lockable frames to protect the prepared hardground
surfaces, Horn Park Quarry, 2012.

shelter, several bespoke wooden padlocked cold-frame-like structures
have been erected to shelter the best areas from weather (and offer further protection against criminal collection). As Sam told me a few
months before the covers were finally installed: 'These will be very
heavy-duty and secure, and offer protection from the weather and
frost, and nobody will be able to get in to rob the fossils again. But
you will be able to come in with students, lift the lids up and have,
eventually, a staircase of pristine surfaces to look at.' Meanwhile, the
British Trust for Conservation Volunteers has been drafted in to
maintain the site annually, spraying weeds and generally keeping it
tidy for visiting scientists who have obtained the appropriate official
permissions, and their apprentices.

Sam locked up and we parted. As he drove off, I felt full of admira-
tion for the way in which the new guardians of Horn Park, at Natural
England and Dorset County Council, were seeking to preserve and

present what remains of this once magical quarry. As Sam had said: 'Covering up a quarry like this one would be like going into the Bodleian Library and ripping up volumes.'

Horn Park is just one among myriad similarly abandoned former quarries in Britain. Because of its critical scientific importance, it has been saved for the nation, but – and this is why its tale is instructive – only by the skin of its teeth. This does not offer much hope for the rest. The Killay Brick Pit had famous connections and some credentials, but at nothing like the level of Horn Park. So it, like the majority of old quarries, has been left to the bine and the flood, forgotten and useless as vegetation encroaches and frost and erosion do the rest.

For all my gratitude for the work they have done, I can only find two cheers for the custodian priesthood who now guard Horn Park, and the lucky few holes in the ground that still command such faithful attendants and interpreters. I can think of no practical alternative; I even admire the didactic purposefulness of both amateur and professional enthusiasts. But a larger part of me regrets the need.

If we still worked our native quarries commercially, none of this would be necessary; and for that reason I applaud Sam Scriven's so far unsuccessful attempts to persuade conservation bodies to repair drystone walls using rocks dug from new or reopened quarries nearby. Because, once a quarry ceases to be worked, one of three things can happen. Either the site is built over or infilled, or it returns to nature and vanishes into the background like Killay Brick Pit, or, as here at Horn Park, it is deemed too precious to be touched. If the latter, then since there is never the money to do any more, the site finds itself prepped and cleaned and covered before disappearing behind fence and padlock. Well-meaning academics spread their pall, and the rigor mortis of amenitization sets in.

Visiting mines, whether working or disused, can never be a practical option except under the highly controlled circumstances of a mining museum, of which several now exist. But these do little to address for most people the central problem with the way we organize our minerals industry. Namely, nobody knows where anything comes from any more because they no longer see it happening. This is only enhanced when mining becomes the preserve of museologists. The neatly presented narratives into which, for public consumption, all things must be shrink-wrapped merely reinforce the mistaken notion that mining is a thing of the past.

3

Mine (dis.)

'But it will roll down on top of us,' I said. 'In time to come, I suppose,' he said, but not taking much notice. 'Years, yet.'

Richard Llewellyn, *How Green Was My Valley*, 1939

On 15 September 2011, a piece of news carried the people of Britain back into the nineteenth century. In an almost precise re-enactment of the 1872 Killay accident described by Francis Kilvert, four colliers had become trapped in a mine in Wales. Three others escaped. Early reports suggested that water had flooded in, probably from old workings; there had also been a rockfall. Fifty rescuers swiftly converged on the Gleision drift mine, a small artisanal colliery operated by just seven men under a company called MNS Mining Ltd. Next day, the missing four – Philip Hill, 44, Charles Breslin, 62, David Powell, 50 and Garry Jenkins, 39 – were found dead.

Such news is always shocking, but this doubly so because these days such accidents happen in places like China, Chile, Russia, Ukraine or the United States, not the UK. Could it be true that British men still went underground to hew coal by hand, in stalls, using blast, pick and shovel, lying on their sides for hours on end in the gloom? Most British people think that coal mining of this sort went decades ago, and that even big modern pits, using longwall mining, mechanical cutters and automatic walking roof-jacks, had died out too, killed off by Margaret Thatcher.

A superficial glance at the figures backs that claim. British annual coal production peaked before the First World War at 287 million tons, which was also before the mines were mechanized, and so was achieved almost entirely by muscle power. This represents a huge workforce. When the major collieries were nationalized in 1947, coal was the largest single employer in the country, employing over 800,000 personnel. This figure grew during the post-war years. For a period, mining employed over a million people. Today the number is less than 3,000. After enjoying its brief post-war boom, coal output went into almost continuous decline – and yet, as a measure of how dominant it had been, coal did not lose its place as the UK's main power source until 1971, when petroleum overtook it. Nor was that its last gasp, for Britain remained a net coal exporter a decade later.

Since the start of the Industrial Revolution – let us say 1760 – something like 26 billion tonnes of British coal have been brought to surface; most of it (all but 773 million tonnes) from deep mines. In 2005, the UK was producing about 20 million tonnes a year, just under half coming from deep mines, ten in England and three in Wales. By 2010 the number of major deep mines had fallen to six. None remained in the South Wales Coalfield. South Wales coal is mined close enough to steelworks for these to take much of the output, but the majority of the 20 million tonnes mined in the UK today is used to generate electricity. Putting that in true context, home-mined coal makes up only ten per cent of the UK generating industry's required supply. The rest of it – 41 million tonnes a year – now comes from Eastern Europe.

The Gleision drift mine was a double throwback because this was a small, private mine too recently opened (and would have been too

small in any case) to be included in the post-war nationalization programme. It was one of just over 90 such artisanal mines still operating
in Wales at the start of the 1990s. By 1997 this number had shrunk to
19 – and to only three or four at the time of the accident. Its history
is obscure, but it seems Gleision had been producing on and off since
the early 1960s, on a steep, wooded hillside high on the eastern flanks
of the Swansea Valley in a scattered rural community of villages,
some of which are very old.

The nearest village, Cilybebyll, is an ancient settlement, part of a
fifteenth-century estate that once encompassed the entire area; a
farming community nestling in a distinctive saddle in the east wall
of the Swansea Valley known as the Alltwen Gap. This feature,
much pointed at by leaders of geology and geography field trips, is
known as a 'wind gap', for it is all that remains of what was once the
bed of another ancient, vanished river that flowed south-eastwards
across the coalfield, long before the last Ice Age even began. This
river and others like it bore little relation to the grain of the land
underlying them, probably because they formed on a cover of much
younger rocks, now long eroded away. But when a glacier gouged out
Swansea Valley, the new and over-deepened course that it cut
followed deep-seated cracks in the revealed basement – and especially a zone of complex and ancient geological faults grouped
together as the Swansea Valley Disturbance. This ran more or less at
right angles to the old, preglacial river courses. After the ice melted,
a new river, today's River Tawe, captured the old river's headwaters –
the Upper Clydach River. Today this tributary, far more ancient than
its parent, thunders down a series of cascades on the western side of
the Tawe Valley and joins with the newcomer at the little town of
Pontardawe.

Cilybebyll, sitting in its abandoned riverbed, seems curiously
remote and tranquil. It centres on a squat little church, St John the

Baptist, with an ancient square grey battlemented tower of limestone onto which has been grafted a Victorian nave of brown Pennant Sandstone. Shaded by cypresses, it stands just off the single-track Church Lane by a small green scattered in spring with a few random daffodils. It does not look like a typical mining village, and it isn't. Big mining still continues hereabouts, however, at several opencast workings, all of which are far away from habitation and perched high on surrounding mountains where almost nobody ever ventures.

As with all modern extraction sites, considerable effort and skill is applied in quarry design to keep the workings as invisible as they can be, out of concern as much for public safety as conventional aesthetics. Even a quarry's afterlife may be planned to deny its former existence. Original land profiles are restored, or the scar modified to mimic the natural erosional features of the surroundings. This is done with the best possible intentions, and quarry designers (many of whom are my friends) are proud of their skills. But I find the extractive industries' self-effacement to be lamentable – it only serves to reinforce the general feeling that quarrying and mining are dispensable activities.

The story of how the Swansea Valley developed during the Ice Age could be repeated for the neighbouring Neath Valley, which runs more or less parallel to it a few miles to the east and whose river also empties into Swansea Bay. A drive up Neath Valley today is a very different experience from the one I remember as a boy. The *Two Ronnies* joke about every town having a bypass so you can drive all day without getting anywhere could have been written with many of the South Wales valleys in mind. Dual carriageways isolate the driver from bypassed communities. A screen of trees, which has grown up along the embankments, obscures places whose existence is hinted at only on roundabout signs. Old spoil heaps have been cleared, and the great conveyors that once ran in long, covered roadways, their snake-like corrugated-iron sheds painted black and perched high above the valley floor on steel or reinforced concrete stilts, have gone too. Forestry Commission plantations clothe the hillsides, hiding the old scars of industry. Honey buzzard and red kite soar high above.

This new landscape more closely resembles that which greeted the first human colonizers of postglacial Britain, than the one which early industrialists would have seen in the late eighteenth and nineteenth centuries. In seeking to return the landscape to an imagined pre-industrial state, we have almost succeeded in resurrecting the prehuman one. This is not a coincidence. A pre-industrial landscape *is* prehuman. It is as though the land has somehow grown younger as it has aged. A new term has emerged for this process: 'rewilding'.

Today, defunct opencast mine workings are sometimes replaced with artificial re-creations of 'natural' geomorphologies deemed right for the location by expert landscape designers and archaeologists. I suppose one could call this 'geomorphological rewilding', creating a 'stage-set' landscape and imposing a new dimension of artificiality on what is already artificial. The concept of wilderness, in Britain at least, is little more than romantic nonsense.

The Neath Valley is the most direct route from Swansea Bay to Merthyr Tydfil, once iron-making capital of the world, where my great-great-grandfather, George Bowen, and his English wife Mary Holloway went to start a new life in the late 1840s, leaving their Devon home and sailing back across the Bristol Channel to where George's ancestors had been born. By then the old forests were long gone, thanks to the early iron smelters that depended on the intense heat of the charcoal furnace. When the wood ran out, technology came to the rescue, developing a method of using coal to smelt the ironstone that – as though by the grace of God – came with it, and which South Wales's first major mines were sunk to exploit. Limestones, cropping out around the coalfield, provided the necessary flux for the furnaces, while sugary quartzites of the Millstone Grit were exploited for making abrasives and refractory furnace-lining bricks. Hundreds of coal mines were sunk over the next century – often in remote, blind-ending valleys where only sheep farmers and their dogs had wandered before.

I had driven to Cilybebyll after the disaster for an article I was writing about artisanal mining. Feeling I had seen enough, I walked back to the car. It was the first time I had found myself near the heads of the South Wales valleys for many years, and the changes that had taken place during my two, perhaps three decades of absence meant that I hardly recognized the place. I began to wonder about other valleys that I had once known well. How had they fared? A plan formed in my mind. The day was still young. Not far away, just south of the South Wales Coalfield's northern ramparts in the floor of a neighbouring valley stood a certain small mining town that I have always regarded as my ancestral home. My mother was born there; but it is not an easy place for anyone to visit, and I had not paid my respects for many years.

I felt a solemn duty drawing me, though there was a more personal motive. Despite the fact that we were by this time approaching our Silver Wedding, my wife Fabienne – who had been waiting patiently in the car – had never been there; never made the pilgrimage to the shrine dedicated to my great-grandfather, George and Mary Bowen's son William. Many times Fabienne had heard me speak in reverential tones of William Bowen. This, I said as I got into the car, was her chance to see for herself the place where he had lived and worked and died; to see the chapel of which he was, to quote the cover of his funeral programme, *'Diacon fyddlawn a Thrysorydd yr Eglwys addiar cychwyniad yr achos yn "SMYRNA"'* ('faithful deacon and founding treasurer of Smyrna').

William Bowen, Colliery Undermanager, and his wife Margaret (centre) with his brothers and sisters-in-law. The picture was taken before the front parlour window of 'Maes-y-Bryn' from which, about four years later, my mother watched his funeral cortege.

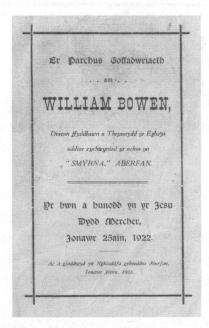

'In respectful remembrance of William Bowen, faithful deacon and founding treasurer of "Smyrna", Aberfan. Fell asleep in Jesus Wednesday, January 25, 1922. And who was buried in the public cemetery, Aberfan, January 30, 1922.'

Smyrna Chapel was not the only monument on the itinerary. I suggested we visit William's magnificent tomb, in buff sandstone and grey granite, and, on a bend in the River Taff, the now empty, grassgreen site of the mighty mine in which he had risen from boy to collier, fireman, overman and undermanager in the great age of industry these valleys and this country knew. It was a chance to see where the man had lived who once owned those candlesticks and lustre jugs which now encumber our home, and with which I cannot part. Most of all, it was a chance to see what now remains of the world that surrounded the first of my ancestors to become qualified in anything. For William had become a mining engineer, a trade closely allied to my own former trade of geologist. I still possess his various certificates, charting his professional ascension and culminating in a letter from the Secretary of State. What is more, he did all that despite having first gone to work down the pit when he was only eleven years old.

Fabienne was less than thrilled at my suggestion, but indulged me. We joined the Heads of the Valleys Road that, in pre-M4 days, my father regarded as his fastest escape route to England. As we drove, Fabienne asked about the chapel's unusual name, just as I had asked him when I first gazed upon the building in 1968. At that time I knew no Greek at all, except for the sound of it; my autodidact father having practised declaiming the *Iliad*, *Odyssey* and bits of a Greek Testament to me in my cradle (reasoning that it hardly mattered to a baby what language was spoken to it). 'Smyrna' sounded a bit Greek to me, so I waited for an answer in a state of greater hope than if I had asked about one in Welsh or Hebrew, like 'Caersalem' or 'Zion'.

My rational, English, classically inclined scientist-technician father, who was forty-eight then, dredged up the fact, which I now repeated, that Smyrna was a port in Asia Minor claiming to have been the birthplace of Homer. (It has become the modern-day Turkish city of Izmir.) This had not seemed, to my twelve-year-old self, much of an explanation for why a Welsh Baptist chapel should bear its name; but after all those years it was still the only explanation I had. I had not read the passage in the Book of Revelation which refers to the mystery of the seven stars and the seven candlesticks, and the seven churches of Asia, to one of which, Smyrna, the Lord sent an angel to say: 'I know thy works, and tribulation and poverty (but thou art rich) . . . fear none of those things that thou shalt suffer.'

We skirted Merthyr Tydfil, now marooned within its enclosing triangle of dual carriageways, and headed south down the valley of the Taff, through the village of Troedyrhiw, where William was born in 1856, exactly a century before me, a coincidence that, together with our both having one Welsh and one English parent, binds me still more closely to his memory. Yet another curious coincidence that extends throughout my maternal line is geology. For I, my mother, her father and his father were all born and raised (and in all cases bar my own, also died) above the very same bedrock of Pennant Sandstone: a rock sequence laid down in a geologically brief interval between the formation of two coal seams – known where I come from as the Penlan Seam (below) and the Hughes Vein (on top).

Usually a few hundred feet thick, the Pennant was deposited between 311.7 million and 303.9 million years ago, a 7.8-million-year slice of Earth history that geologists have named the Westphalian Stage of the Carboniferous Period. Britain then lay somewhere near the equator and was covered by a system of switching, braided rivers that laid down the many sandstones separating the seams that give the name of 'coal measures' to this section of the geological column. Despite the fact that we three generations who lived on these rocks did not all live in exactly the same place (rocks of any age, once folded and faulted by Earth movements and shoved around on the backs of drifting plates, can crop out almost anywhere on the surface of the Earth), none of us wavered from these same few hundred feet of sandstone. We all were raised on Pennant Hills, among their building-stone quarries, ancient adits cut to scrape out minor, ashy coals and the innumerable spring lines that bubble up wherever the porous sand-stone is interrupted by impermeable clays.

Drawing closer to the place, we drove out of Troedyrhiw, ducking under the low railway bridge and crossing to the western side of the valley. Although it had been years, there was not much to remember. The village has one main road running through it, and I knew that Smyrna Chapel would eventually appear on our left – unmissably huge, exquisitely mournful – just before a right fork that led up, past the double-fronted house that William built for himself, his wife Margaret and children Catherine and William Holloway, to high Bryntaff Cemetery where he, she, my great-aunt, her husband (and two uncommemorated sons, as I was to discover) now rest in the same triple plot. But as we drove along the slight curve of Aberfan Road, inconceivably it seemed that Smyrna had vanished. Nothing seemed to make sense. I pulled over into a convenient car park that I didn't remember.

We sat there, the car engine still running, as I tried to get my bearings. For some reason, as I searched for familiar buildings, my thoughts drifted back to that gloomy pre-Christmas day in the quarry behind Hendrefoilan House, after fighting my way in to find myself staring at the apparently horizontal bedding of the rocks in the quarry face, abandoned a century and a half earlier. The Pennant, at its thick-est under my boyhood's native soil, finds itself standing on edge there, dipping steeply towards the heart of the coalfield. In the interior, as here in Aberfan, although less thick, the Pennant *looks* thicker because it is lying at much lower angles. Here, instead of presenting its broken ends to the sky in sudden ridges, the near-horizontal Pennant gives rise to broad, soggy uplands – where the streams rise and under whose tussocky moors my maternal ancestor hewed the coal.

On that unhappy day, I had closed my eyes and heard the chink of quarrymen's hammers, the back-and-forth rasp of rock saws, the creak of carts and the jingle of horses in their traces. It was 1855. Lewis Llewelyn Dillwyn's great house was slowly rising from its foundations. One year earlier, my great-great-grandfather and his wife Mary had passed through Swansea on their way to Merthyr from their home on Exmoor in Devon. She was about to give birth to my great-grandfather, William Bowen, on these very rocks here on the far side of the coalfield.

I had also seen the shifting, sand-laden rivers of 300 million years ago, draining the Wales-Brabant Island, whose eroded roots today form the mountains of Mid Wales. I felt the compression that converted this sinking trough of gelatinous sediment into a structural downfold that would eventually form part of a new mountain range, as the supercontinent Pangaea came together, uniting every continent on Earth and stretching its deserts across all latitudes.

I had imagined diving into one of these bedding planes and following it – like my miner ancestors, keeping to the same stratigraphic level – driving my road through rocks of the same age as those in front of me. I followed the remains of a hidden landscape, buried for 300 million years; the river systems laying down their rippled sands, occasional jungle patches in-between the levees, impenetrable tangles of fallen, undecaying trunks sinking slowly into the thick peat, a future coal bed; festering lagoons floored by black, oxygen-poor, carbon-rich muds that would one day make the shales used to form the almost unbearably red bricks of Dalton and raise the black spoil mountains that have not yet grown above the head of my great-grandfather's mining village, smothering the tough hags, burying the subversive springs that dot the hillside and do not cease to bleed their waters into the growing heaps, never stopping, meaning trouble.

I drove my imaginary roadway through that landscape's remains, diving down the steep dips before me in the quarry wall, then, following the bedding as it levelled, turning east, circling around the northern hinterland of Swansea; passing Neath and Tonypandy before bending north somewhere near Abercynon. There, in the interior, where the rocks lay mostly at or near horizontal, I picked up a vein of coal: the Brithdir Seam. Hardly present at all near Swansea, here in the north-east where coal lies thick the forest had grown long and strong and built a noble seam, often a yard top to bottom, which was exploited in many mines.

Eventually, nearing the surface, I saw myself exit the Brithdir Seam and, burrowing through a veneer of glacial debris, emerge onto a hillside overlooking the village where William Bowen lived and died. I look about me. Two pithead winding gears stand below, on an area of flat land across the River Taff, where that Welshest of rivers, on its journey to Cardiff, swings west within the confines of its glaciated valley. Cloud sits low over the mountains, closing like a lid over the greasy streets. A road rises obliquely towards me (called Station Hill because near where I stand, it crosses one of the valley's two railways). A hearse, drawn by a pair of black horses with cockades, is just pulling away from Smyrna Chapel for the short journey up the incline, over railway and canal, and then up to the hillside cemetery of Bryntaff. The horses' breath hangs in small clouds on the still air. Mourners, gentlemen only, trail behind, droplets condensing on their black wool coats. It is 30 January 1922.

As the cortège passes a detached, double-fronted house in Pennant Sandstone and brick called Maes-y-Bryn, my mother, just turned two, peeps between the heavy drawn curtains. She watches her father following the cortège, staring hollow-eyed at the coffin and its brass fittings, arrayed in all the grandeur of grief. He does not see her. Gentle, kind Holly Bowen, with his belly scars from a Turkish sniper bullet, thinks of his forceful, fearsome black-eyed father, dead five days of 'bronchitis' and soon to lie, at a select location, in a brick vault cut into the hillside's veneer of boulder clay beneath a grey granite obelisk that for the moment bears only the name of his wife, Margaret, dead two years before.

My mother stands in the house where she was born, the house her grandfather built, on rocks exactly the same age as those underlying a patch of land near Swansea where a woman called Nettie is just then nearing the end of her life in a rubblestone cottage. In only another four years, Nettie's market garden will be sold, and on part of it, an incongruous suburban villa will rise.

'Well?' Fabienne asked, bringing me round, unaware of the daydream that had been going on inside my head. I turned off the engine, and in the silence it dawned on me that we were actually sitting where Smyrna had once stood. Where the congregation at

Smyrna Chapel in 1984.

Smyrna Chapel in 2012.

William's memorial service sang 'Moriah' in 1922, local residents may now leave their motors safely off-road. The car was standing almost exactly where he had lain in his coffin, at the head of Smyrna's centre aisle. Merthyr Tydfil Borough Council really had 'paved paradise and put up a parking lot'.

I got out and walked around. Sure enough, the more modestly proportioned *old* Smyrna Chapel was still standing. This 1877 original became the Sunday school when its now-demolished 1901 outgrowth sprouted at right angles from its south side. It had been lovingly restored and cleaned and converted into a 'community facility'. It was locked. I noticed a telephone number in the window and dialled it. Had anyone answered, I might have asked 'what had they done with my great-grandfather's chapel', so perhaps it was as well nobody did.

*

It was Good Friday 1968 when – after paying our respects at the family sepulchre – my mother, grandmother and I first walked along a row of pristine gravestones in Bryntaff Cemetery. Everyone buried there had died on the same day – 21 October 1966, eighteen months before; a day on which I, like most of these dead, had been at school, only in my case it was forty miles away on another hillside. Towards the end of that morning, our teacher, his full face pale and visibly shocked, called for attention and announced that a terrible disaster had occurred, and that hundreds of our exact contemporaries were lying trapped beneath a landslide. Of my classmates at that moment, I was probably the only one ever to have heard of Aberfan.

As I sit at home in London, forty-five years later, two heavy brass candlesticks, grown dull and buttery for lack of attention, look reproachfully at me from the mantel. It is perhaps more than a little pitiful to be made guilty by inanimate objects; but nevertheless I get out the Brasso. It is not the candlesticks, I tell myself. It is the ancestral voices.

Sculptor Auguste Rodin, according to Anatole France, once said: 'All the relics of the past talk . . . they murmur to us a hundred touching confidences about the honest men who fashioned them.' I would add, 'the men who owned them and the honest women who cleaned them'. But these candlesticks were not always the precious objects they seem to me today. Once, they lit the parlour of a mean cottage in Troedyrhiw where William was born, and later, Nixonville; dwellings older and even shoddier than my late Victorian north London house.

I wonder how much polish was lavished on them then. Over-vigorous digging of dribbled wax from their cups has left deep gouge marks, which suggests that they were not always revered as I revere them. I often wonder whose hands made those marks. I can never know. But I remember how my grandmother would intone, as she applied the Brasso: 'These were in *Aberfan*, you know.' They had been part of the establishment of William Bowen, a name she, although a Llewellyn from the rural west, would never utter other than in tones of awe. His aura reflected upon them a glory that vastly outshone all those candlesticks' other owners, previous or subsequent.

She repeated those words each time she handled anything from that time. All had 'been in Aberfan'; and they connect me to what I saw then as a Homeric age of men who delved in the Earth with a mythic grandeur and a manly dignity befitting heroes, unabashed to deal at first hand with God Himself. The name Aberfan, known to me long before my first visit, rang in my young ears like 'Camelot', 'Athens' or 'Ithaca', or even the name of some infinitely distant geological epoch; echo of a time before men learnt fear, before life grew small around them; before these few household items, with their tattered legends, became its last surviving talismans. It seems absurd now, but hearing this familiar name on that drab school morning was like learning of the sacking of Jerusalem.

Later, watching the news (Aberfan's was the first man-made disaster of the television age, just as the eruption of Krakatoa in 1883 was the first natural disaster of the telegraph age) I saw the death toll mount. It was about then that a thought occurred to my grandmother, one that would resurface again and again until she died; namely, that just as her husband had been saved from almost certain death in France by volunteering early and fighting in Palestine, I had been saved from Aberfan's catastrophe thanks solely to their decision to sell up and raise my mother in comparatively genteel Swansea. Perhaps it is a grandmother's prerogative to claim oracular powers. Anyway, she henceforth claimed full credit for my 'survival'. We read the freshly hewn names that Good Friday, and she said again, as I knew she would: '*You* should be here, by rights, you know.' Not surprisingly, I appear in photographs to be sticking closely to my mother.

I remember the day's cold spring light, the violent yellow explosions of the daffodils on each grave; and despite the almost inexpressible air of tragedy that hung about us, of being incongruously excited by the sight of the spoil heaps, high on the skyline, the tallest I had ever seen; and by the thought of what it would be like to stand on the top of one of them. I was excited too by the thought of all that volume of rock – 2.66 million cubic yards, the Disaster Tribunal later estimated – having once lain 1,500 feet or so below the valley floor, and which was now a thousand feet above our heads. I thought of the voids it must have left behind, the long, dark timbered roadways that my great-grandfather had known and helped to drive. And I remember being torn between an old, heroic pride (manifested in William's triple plot and great obelisk) and this paralysing overburden of grief that pressed down everywhere in Bryntaff Cemetery.

Back in 1922, after William Bowen's funeral, probate ground dully on to its conclusion. Maes-y-Bryn was sold, and my mother and her parents left Aberfan for a new life. There, during the next war, she would meet a young industrial chemist, come from Manchester to work as a spectroscopy technician in the city's then extensive metallurgical industries, and would walk with him on the Pennant Hills above Hendrefoilan House. There they made plans to live, when they could afford it, on the edge of the countryside, near to a school. That time came in 1957, and my mother would live in that same house for exactly fifty years before leaving it for the last time aged eighty-seven, on a summer morning, for a hairdresser's appointment she would never keep. My father lingered on alone for four more years before he too would leave that house for ever, in a dressing gown at four in the morning, in the rain.

When I had stood in Hendrefoilan Quarry, the bad day which her-alded my father's eventual transfer to a nursing home still lay a week in my future. I feared there was worse to come, replaying in my mind an incident of the evening before, when my father had turned to me with chilling politeness, called me 'sir' and asked my name. The great forgetting, which eventually claims all things, had its toe in the door. The river of time was wearing away; once-remembered hills were being forgotten, grain by grain.

Still reeling slightly, I drove out of the new car park and past Maes-y-Bryn. A small group of lads in baseball caps was sitting on its steps; steps before which a throng of several hundred miners had stood in April 1912, listening to William with his ardent eyes and great black moustache like a breaking wave. The men had been on strike since February, as part of the Mining Federation of Great Britain's cam-paign for a national minimum wage.

Maes-y-Bryn in 1984; the house William Bowen built, where my mother was born, and from whose front railing William urged the strikers back to work in 1912.

The strike had come after two years of riot and unrest in South Wales, so feelings were high. Receiving news that a settlement was near, and of the Government's pledge to introduce a Coal Mines (Minimum Wage) Bill, the men found themselves unsure. The Federation had said they should go back to work. But what was the Federation? They weren't *from round here*; they weren't *village*. The men decided they should 'ask Overman Bowen what *he* thought'. And from the steps of Maes-y-Bryn, William told them to listen to their Federation and put an end to their sufferings. And not only theirs; for during those long strike weeks, William and the other nine overmen had worked night and day to keep the mine drained, ventilated and safe – ensuring that there would *be* work to return to when hostilities ceased. Cease they did, the very next day, 6 April.

The colliery took a team photograph of the ten men who had kept

Merthyr Vale Overmen on 6 April 1912, the day peace broke out in the South Wales Coalfield. William and his colleagues had kept the mine drained and ventilated during the 37-day strike. To the right of William (back row, second from left) is his contemporary William Thomas.

the mine alive. They stare out of it, serious, hollow-eyed from their labours. The end of the strike was big news, but it was not on the front page for long. Ten days after that photograph was taken, the liner *Titanic*, its maiden voyage delayed by lack of fuel, its top speed cut by the need to conserve it, overcrowded because so many other ships' sailings had been cancelled, slipped beneath the Atlantic with its remaining stores of Welsh coal.

I drove up the diagonal of Station Hill, switching back hard right over the old railway and canal (now a cycleway called the 'Taff Trail') through the cemetery gates and pulled up outside its chapel of rest. From there, Fabienne and I walked up to the broad, level road on which William's monument stood – and where for the second time that day I found myself looking about in confusion. Several tonnes (it turned out later to be four tonnes) of memorial masonry had apparently disappeared.

As I got my bearings, I remembered from photographs taken on previous visits that the headstone to the right of William's tomb had been made of brilliantly white Italian marble. I could find no such stone now. In a strange reversal, the cemetery that once shone with the colours of many exotic rock-types has faded to a dull mousy grey, recalling the coating of black dust that, when the mine was working, used to coat everything in the village.

Closer inspection revealed that what I had taken for a grey-black headstone was, in fact, cut from the dazzling marble of Carrara, its whiteness now concealed beneath an ugly patina of grime and lichen. The bearers of the barely legible names inscribed upon it probably had no relatives left alive – or else, like me, they were far away. I turned my gaze slowly to its neighbouring triple plot.

Standing in the middle, overgrown on three sides by ivy, was a sandstone box tomb. Could it have been the base on which the rest of William's monument had once stood? I parted the undergrowth and

William Bowen's tomb in 2012, after being dismantled by
Merthyr Tydfil Council for 'failing a stability test'.

discovered a shrouded urn carved from grey granite – just like the one
I remembered. Next to emerge was a square capstone in the same
material. And on the far side of the box tomb lay the recumbent
obelisk, on top of which the other two elements had once sat.

The obelisk's uppermost face was blank, as were its sides, the
inscription unfortunately – *inexcusably*, I felt – turned to the ground.
My last hope of making a positive identification lay in a younger
memorial to William's daughter, just to the left of the box tomb.
Another memorial was certainly present there; I pushed back the ivy
until I could make out the name Catherine Lewis – my great-aunt.

Silently contemplating the wreck, I wondered how such neglect
could have come to pass. The original receipt for the plot, Number
1/17 (preserved in the family archive), records that William had paid
an extra guinea for his 'select location' right on the cemetery's main
drag. Now it looked more like a crime scene. Four fence-posts had

been hammered in at its corners, from which fluttered a few tatters of sun-brittled plastic warning tape. A notice tacked to one of the posts read: *This memorial has failed a regulatory stability test and is unsafe. It has therefore been temporarily supported/dismantled. Please contact the Cemetery Office for advice.*

This was no particular insult – no more than the general insult that time metes out to everyone and everything. I discovered afterwards that Smyrna Chapel had lain derelict for years – despite a £50 bequest from Aunt Catherine – and become an even greater eyesore than it arguably was already. And so the extension for which William had led the fundraising and which had swelled Smyrna's capacity to 600 singing souls, had been demolished to general approbation in 2002, a century and a year after first opening its doors.

Times had also changed for William's memorial. Cemeteries like Bryntaff were designed as places for quiet respect and contemplation. In 1922, a few dabs of mortar and the force of gravity were considered sufficient to hold four tonnes of rock together. Memorial masons never imagined that adolescents might climb up their creations and swing from them. Nor that any such youth might argue that his injuries deserved compensation. Herein lay the reason the local authority had dismantled the tomb.

The dismantling seemed to point to that bigger loss, happening everywhere in the British landscape: the gradual and often deliberate effacing of all traces of the industrial exploitation of the rocks beneath us. Mining and quarrying, despite being fundamental to the way we live, are banished, leaving us in a strange, unreal bubble of privilege, receiving the benefits of raw materials and energy but clueless about their origins.

Growing up as I did on the edges of a great coalfield, mining activity past and present had forced itself unavoidably on my attention. All of Britain, not just its more economically important areas, was then pockmarked by hundreds of thousands of mines, quarries, brick pits and spoil heaps all proclaiming our debt to the past in general, and, especially, our *local* past. The bones of the Earth were everywhere laid bare.

As I knew from Nettie's cottage, my parents' house and those that were built around it later, not so long ago all buildings were made from materials occurring on site or close by. Today, if a planning authority were to rule that any new building must use local materials, it would end up costing a lot more. The old logic of sinew and muscle has been turned on its head. Broadly speaking, the more distantly Earth materials originate, the cheaper they are – and this has mostly happened in my lifetime.

Turning our backs on the colossal wreck of William's once-proud monument, we climbed to the top of the graveyard to discharge the main and more serious purpose of our visit. White marble arcades now link the graves of 116 of my contemporaries, who died together on the same day in 1966 in Britain's worst ever mining-related disaster, when a spoil heap perched above the village on the slopes of the Merthyr Mountain slipped down, engulfing farms, houses and the village school. On that day, the price of what heedless men had done (long after William Bowen was dead and buried, I am happy to say) was paid by miners' children. The arrogant men who had come to run the Merthyr Vale Colliery had ignored what ordinary people knew well about the slopes; ignored what the geology of Mynydd Merthyr was shouting at them. This refusal to hear what the rocks are saying is happening, to all of us now, on a global scale.

As we left Aberfan, I felt that I owed it to William Bowen to restore his monument and, symbolically at least, repay the debt I personally owe to the miners, quarrymen, masons and foundry men who inhabit my family tree, who strove all their lives to find a better life for their children; men and women whose names are too often turned to the ground, as though in shame, as if the past and all its pain can ever be simply tidied away, dismantled, remediated, converted into a recreational amenity and made safe.

Minerals – another way of saying 'Earth materials' – can only be won where they occur. Therefore they often give rise to communities where none previously existed. Elsewhere in the world, these settlements would be called townships. In 2002 I went on a writing assignment to the Rössing uranium mine in Namibia. Sited about sixty-five kilometres inland from the quaint, clapperboard seaside town of Swakopmund, the Rössing open pit is one of the world's biggest holes in the ground, three kilometres long, half as wide and well over 300 metres deep, mining the world's largest granite-hosted commercial uranium ore body.

The body lay undiscovered amid the barren wilderness of gravel plains and rocky outcrops of Namibia's Erongo Region until 1928, when a prospector called Captain Peter Louw tested a black mineral sample he had picked up about twenty kilometres west of the present mine. He and others made several unsuccessful attempts to generate interest in the deposit, and it was not until the 1960s that a subsidiary of what is now Rio Tinto plc, the mining giant, took up the option. After further evaluation RTZ (as it then was) gave the mine the go-ahead in 1973. The mine and its refining plant were designed to produce up to 4,500 tonnes of uranium oxide every year, and after beginning operations in 1976, reached full production three years later in 1979. By the time I visited, the mine was the fifth largest producer in the world, accounting for eight per cent of global production and responsible for ten per cent of Namibia's total exports. All this was achieved in spite of the fact that the ore was very low grade – containing less uranium per unit weight than many mines' tailings (as was more than once proudly pointed out to me).

But mines and processing plants need a workforce, and a workforce needs nearby accommodation. Namibia is a vast and mostly empty country, and at Rössing there simply was nowhere 'nearby'. The mine had to build its own town, which it did – and they called it Arandis. Just over 7,500 people live in the neat settlement about fifteen kilometres north-east of the pit where they all work, on the other side of the main B2 highway out of Swakopmund. Irrigation greens the roundabout on the approach road, and a few lawns set around strategic amenities; but Arandis is a utilitarian place. Like the desert in which it sits, it is mostly silent, gently rocked now and then by the boom of charges being let off in the pit. At the roadsides and in the parched gardens, vicious Namibian desert grass bleaches under the numbing sun, the barbed seed-cases lodging and drilling into your boots until finally pricking your feet months later, to remind you of your visit.

As I drove around the town in my little hired Kombi, it struck me that Arandis exists for precisely the same reason as Aberfan. It was put more or less in the middle of nowhere by a mining company needing housing for its workers. I could even detect a parallel between the little fern-leaf housing estates with neat culs-de-sac of identikit dwellings erected for the Namibian miners, and the basic, unadorned ribbon frontage of Nixonville's brick terrace (now rendered and restored to within an inch of its life) where my grandfather was born. I wonder, though, whether Aberfan's nineteenth-century community leaders ever met to consider attracting inward investment and diversifying the town's economic base away from its single great employer. Such ideas occupy the minds of Arandis's luminaries today: people like the Rössing Foundation's director, Job Tjiho, and Florida Husselmann, the town's 'chief executive'. I doubt their nineteenth-century Welsh counterparts ever looked further, in this world, than King Coal. For everything else, there was Divine Providence.

The spectre that hangs over mining towns is the same everywhere. Once the resource is worked out, or has become uneconomic, what remains is a remote settlement with no reason to exist where it does. In frontier societies, and former ages, such settlements would simply be abandoned and allowed to sink back into the Earth, and I have visited many of these in the former silver-mining districts of Nevada, complete with tumbleweeds. The British countryside is peppered with failed settlements, whose traces in the forgotten landscape lie under the soil like chickenpox scars on an old face, or the footprint of furniture pressed into the carpet of a cleared house. These failed villages generally died centuries ago, during crises such as the Black Death, and have often been completely forgotten – unless, sometimes, one sole edifice still stands as its memorial, usually the church, the only stone building.

Communities in an agricultural economy are potentially long-lived because agriculture is perpetual. Mineral exploitation, by contrast, is always a temporary use of land. Yet in modern Britain, settlements are rarely if ever abandoned. In our age, property owning binds us into an iron triangle of mortgage, marriage and work; and there is more to it than that. All mineral exploitation may be temporary, but coal mining often went on for over a century, and the villages that grew up around it had time to become more like the settlements characteristic of farm economies.

Such villages operate to a large extent as autonomous societies, often hostile to outsiders, from whom nothing is asked except that they stay away. Coalfield mining villages were given the time, very often, to develop a resemblance to the medieval 'peasant' village described by the social historian Henry S. Bennett, who wrote, in his classic 1937 work *Life on the English Manor*: 'Their village was their world: beyond, some ten or twenty miles led to the local shrine or great fair, and there, perhaps, once or twice a year they made their way; but, for the most part, their lives were ground out in a perpetual round from one field to another, and so to the next ...'. And when Russian sociologist Teodor Shanin wrote of how a peasant community's emphasis on family and household typically leads 'towards segmentation into units of high similarity and low mutual interaction', he could have been describing village life in the nineteenth and early twentieth-century South Wales Coalfield.

Many of the smallest and most remote mountain mining hamlets with their ephemeral pits have come and gone; many have been utterly destroyed by later, opencast workings – like the lost hamlet of Pantywaun, north-east of Merthyr Tydfil. Its fifty or so houses, a matching trio of pubs and chapels, school and community hall, have disappeared with the mountainside to which they once clung. But many more places have faded into quiet, listless dormitories, bypassed by the modern world. Where once they exported energy as coal, now they export it as people.

The consequences are familiar. Housing falls in value and becomes hard to sell. Sometimes, nearby larger towns and cities may seek to relocate hard-to-home welfare cases to wherever housing is cheap. Per capita income falls; shops and pubs close. In Aberfan today you do not often meet people walking down the streets – not like you once did, anyway. Those who work mostly do so elsewhere, and travel by car. The parades of small traders that once lined the Aberfan Road have gone. There is a minimarket; the Aberfan Arms, which seems more often closed than open, and a fish and chip shop. All over the village, houses are in good decorative order but their doors are shut, blinds drawn, just like everywhere else and not like of old. Grand old buildings – some chapels, working-men's institutes and the like – have seen their former uses wither. Buddleia has sprouted from their pediments like old men's eyebrows. As with Smyrna, compulsory purchase and demolition orders have repurposed their past glory to newer, more mundane uses. Today, Aberfan is nothing if not tidy and clean, and it takes pleasure in the fact that, for once, it can be so. But to me, the dead hand of municipal tidiness and *aménagement* that is so characteristic of such places hangs heavily in the air.

When the historian A. L. Rowse finally emerged from his Cornish boyhood chrysalis as a flamboyant and outspoken Oxford don, he made a short pilgrimage to nearby Banbury to see the places where his father had worked as an ironstone miner, in temporary exile from Cornwall during the First World War. Today, I approach Aberfan in the same spirit. I walk its streets, as Rowse did Banbury's, 'with the same nostalgic loyalty so strong upon me when it is too late: that strange place which had been the scene of an unknown passage in the life of – for all he had done for me – this unknown man'. For me, the unknown man is William Bowen.

William Bowen (centre, back row) in 1897. Detail from a group photograph of colliery officials of the Nixon Navigation Company Ltd, Merthyr Vale Colliery. Bottom left is the manager, Robert Snape, sitting next to the Agent, Mr H. E. Gray. To the right of William Bowen stands William Thomas, who would also rise to the position of Overman.

Aberfan is an elusive place. Even though it is located in the valley of Wales's national river, connecting Merthyr to the capital, it does not sit on a road to anywhere. It seems as much a cul-de-sac as if it were one of those mining villages lost up some blind-ending cleft in the coalfield's bleak moors, like Glyncorrwg or Gilfach Goch. The old road, which runs down the valley's east side, boasts a signpost to adjacent Merthyr Vale, but none to Aberfan. The new A470, driven down the pine-clothed western side of the valley just upslope from Bryntaff Cemetery, bypasses Aberfan entirely. Only when you peel off to neighbouring Troedyrhiw, William Bowen's birthplace, does the name first appear, on a mean and mildewed fingerpost. Aberfan has become a non-destination. Perhaps it is better this way. While the village appreciates sensitive visitors, the last thing it wants is gawping strangers.

I have spoken already of a photograph, taken on Good Friday 1968 on my first recorded visit to Bryntaff Cemetery one bright spring day, made bleaker and colder by the picture's ageing colours. Trees above the graveyard are bare. Bunches of thick, bright daffodils stand on the fresh tombs, every headstone new. Towering hugely against the pale, cloudless sky stand two enormous black pyramids of spoil, streaked with white. The summit of the tallest, Tip No. 5, reaches 800 feet above where we stand. Tip No. 4, a little lower down the hill, is about 100 feet shorter. Nothing grows on either. But when my great-grandfather's funeral procession passed through Bryntaff's gates on that dull winter day in 1922, not one of these high tips on Mynydd Merthyr was yet born.

In the foreground of another photo, taken on the same day, stand three figures: my grandmother, my mother and between them, aged eleven, me in a bobble hat. Having just gone up to grammar school, I am lately in my first long pants, which are already too short. As usual, in imitation of my father (who took the picture and meticulously noted down its time, date, shutter speed and aperture setting), I am carrying a camera gadget-bag. Forty-six years had passed since my mother stood, aged two, at the heavy velvet curtains, fringed with beige tassels, of Maes-y-Bryn, watching William Bowen's cortège – which is about the same number of years that have passed between the shutter click of my father's camera and now. A few minutes before this photo was taken, we had placed our own flowers on his towering monument.

William Bowen's tomb, Good Friday 1968. My mother cleans the inscription while grandmother conceives the idea that by leaving Aberfan to live in Swansea she and my grandfather saved my life. Tip No. 5 (highest) and Tip No. 4 visible against the skyline. Remains of Tip No. 7 and its slide cover the middle distance below Tip No. 4.

By the time William attained my age in this photograph he was already a working man – operating the brattices, or fire doors, deep underground somewhere in the valley below, possibly at the Duffryn Colliery near Troedyrhiw. At some stage, maybe as early as 1872, he had moved to the Nixon Navigation Company's new mine, just then being sunk at Merthyr Vale – a mine destined to become one of South Wales's most advanced and productive pits, and whose twin pitheads and sheave wheels dominated this valley for over a century; the mine that both gave birth, and brought death, to Aberfan.

By the time his son William Holloway Bowen was born in 1891, the family was living at 26 Nixonville, a company-owned terraced cottage in the shadow of the pit, squeezed between the road and the River Taff. The mine stood on a flat meander loop that gave enough space for its pitheads, boiler and engine houses, mighty ventilating machine, gas works, repairing yard, pit wood depot, timber yards, reservoir, limekilns and rail sidings. Coal could be taken south from there to the great port of Cardiff, across a widened neck of land reclaimed from the diverted river, while waste would travel north to spoil heaps, then confined to the east side of the river. Not until the First World War did tipping commence on the west bank, establishing a bridgehead from which it then began a relentless climb up the slopes of Merthyr Mountain.

Sometime in the early 1870s, two things changed William's life, possibly before he was even into his twenties. The first was the temperance movement, onto whose wagon he jumped (or possibly fell) one night while returning from the pub, never to touch another drop, play cards or smoke until the day he died. He set about improving himself and learning Welsh properly. Although one could hardly be anything other than Welsh with a name like Bowen, his parents George and Mary spoke only English. But soon, William would be able not only to converse with his workmates in their language, but to preach to them in chapel. And with English as his mother tongue, he could just as fluently address the mine owner's agent, Mr Gray, and manager, Mr Snape. In due course, he added French to his languages, and made fact-finding visits to mines in France and Belgium. I have a postcard sent by him from one such trip. Its stamp had been removed, but I found it again, stuck in my mother's childhood album. It bore a Brussels postmark from 1898.

By the age of twenty-one, William had become the founding treasurer of a Welsh Baptist chapel on Aberfan Road, and to which, outside the mine, he would devote nearly all his life. This made him one of Aberfan's more prominent citizens, for not only did he have almost the entire male population's attention for six days a week, but every Sabbath he claimed a sizeable chunk of the rest. As with language, so with religion; William was no stranger to the established Church in which he had been baptized, and where high-ups could be nodded to (as well as his doctrinal bets hedged). William saw the advantage in playing for both teams, insisting that my mother, for example, be baptized in both places, just in case. Sundays in William's household were particularly arduous; everyone who happened to be under his roof that day having to render service unto the Lord at least twice – 'once in good Welsh, and then in perfect English' as the poet Robert Graves might have put it.

Another institution to which William owed much was the Glamorgan Technical Instruction Committee. This and others like it across the country had been set up after the Technical Instruction Act of 1889, which belatedly set out to improve British technical education through a combination of night schools and correspondence. In our self-indulgent sybaritic age with its infantile pleasures, one can only be staggered at the sheer capacity for work that animated nineteenth-century men and women. After his shift, and presumably his fireside bath, William (by then in his forties, the average age at death for a British miner) would head off on foot – initially to Treharris some three miles away (via the short, straight route, over the mountain), returning late at night.

I have his certificates from a two-year course in the Principles of Mining. Each is printed on heavy vellum: durable objects, for damp houses and people unused to paper. The later one reads:

The earliest of William Bowen's professional qualifications, which led eventually to a letter from the Secretary of State certifying him as a qualified under-manager of a mine.

GLAMORGAN COUNTY COUNCIL
TECHNICAL INSTRUCTION COMMITTEE

This is to certify that *William Bowen* diligently attended the *Principles of Mining* Class at *Treharris* during the Session of *1898-9* and at the Annual County Competition was placed in the *First* Class *and was one of the Members of the team which won the third prize of £2-0-0.*

William Hogg, Secretary of the Committee.

In 1900, William next began work on a correspondence course from South Kensington College, London outpost of the City & Guilds, and one of the institutions that was eventually folded into the Royal School of Mines, Imperial College. Locally, things had moved closer to home – no more treks over the mountain. A copy of Pendlebury's *Arithmetic* was presented to him (according to the bookplate inside) by the Merthyr Vale Evening Continuation School, 'for regular attendance' in 1899–90. Finally, among his papers I find a letter from the then Home Secretary, Sir Kenelm Digby, certifying that William Bowen was now qualified to act as the undermanager of a mine anywhere in Her Britannic Majesty's Empire.

An essential part of that licence to practise was the assurance of sobriety – at which point one might discern how William's career plan, in work and outside, fitted together. This visit to Belgium had made William the first of my ancestors to go overseas. This paper made him the first to gain a qualification of any kind – and (as a result, I like to think, of some fault in our shared genetic basement) it happened to be in geology's most closely allied trade of mining engineering.

So, Colliery Overman Bowen moved out from under the shadow of the pitheads to the more salubrious western side of the valley where the village of Aberfan had come into existence during two decades of hurried building. His new house stood just below the canal and railway line running along the flanks of the Merthyr Mountain, just below the gates of the municipal cemetery, next door to the chapel hall (now replaced by a modern house) and close enough to Smyrna to hear the singing.

It was perhaps more satanic mills than Jerusalem, but hope was strong, evident from the chapel founders' choice of name – which, of course, had nothing whatever to do with any association between that ancient city and the heathen Homer. Smyrna, being the second of the

Seven Churches of the Apocalypse (according to the Revelation of St John the Divine), was told by the Prophet: 'Fear none of those things which thou shalt suffer ... be thou faithful unto death, and I will give thee a crown of life.' So far, alas, it didn't seem to be working out too well.

At the best of times, and despite the fierce loyalty they evoke, the mining valleys of the coalfield are challenging places to live; yet those born to it can find it an equal challenge to live anywhere else. Even now, since the black tips have nearly all been cleared and chimney soot and coal dust no longer cover every wall, when the weather closes in, the valleys can be dark, forbidding places. The terraces – not so drab today, thanks to modern paints – line the loud roads, shiny with wet, and leaden clouds obscure the heights that on clear days seem to offer hope of escape. Then, they rear instead like prison walls. Emerging from below ground to such conditions might well have seemed like exchanging one form of burial alive for another.

In the late nineteenth century, mining was not only a highly dangerous occupation but a poorly paid one, an existence punctuated by strikes not over demands for increased wages, but against reductions and privileges withdrawn. Under those circumstances, no father wished the life of a miner upon his son. The dream of every underground worker was that his son might have a regular, clean job, with reasonable hours, good pay and in the Lord's good light (but out of His not-so-good weather) and no heavy lifting. Such, for a Victorian mineworker, was always the most valuable gift he could pass on.

My grandfather was saved from the mine, entering the drapery business before the outbreak of the First World War. When Kitchener's call came, he volunteered early and fought in Palestine – a decision he always believed saved him from the killing fields of France and Belgium where nearly all of his contemporaries from Pant Glas School died. Having survived the war to end wars, thanks to an excess of patriotism and nicotine addiction, Sergeant Lewis-Gun-Instructor

Holly Bowen and my grandmother (made, by marriage, her mother-in-law's exact namesake) moved back into the parental home. There, in Maes-y-Bryn, my mother was born in 1920, among its candlesticks, lustre jugs, Davy lamps and unreadable library editions of Dickens and Scott.

For her first two years, Aberfan was home. But then William the patriarch succumbed according to his death certificate to 'bronchitis', a condition to which he was undoubtedly made vulnerable by the unmentioned dust on his lungs. The young family sold up and moved away, leaving William and Margaret reunited in plot 1/17, while on Merthyr Mountain, the colliery waste climbed higher and higher on the northern horizon.

In the forty-four years that elapsed between William's death and that fateful morning in 1966, many things in mining had changed; many others, not so much. Merthyr Vale Colliery had been sold to the Powell Duffryn Company in the 1930s, and since New Year's Day 1947 had been vested, with radiant but ill-founded hope, in a new National Coal Board – an 'obdurate and ignorant monolith' as it was later described, whose rule proved no less arbitrary or despotic than that of the coal kings, and whose arrogance and incompetence was laid bare by the inquiry into events that made Aberfan's name for ever synonymous with disaster.

Coal-winning at Merthyr Vale had also produced millions of tons of black shale, as well as waste from the coal preparation plant and ash from the boiler house (some of it still hot). This mixture was then loaded into drams and hauled by stationary winches in 'journeys' along a mineral line. Ever since tipping had begun west of the Taff, the waste crossed the river and Aberfan Road on a rising iron bridge, trundling high above the roofs of terraces at Ynys-y-Gored, along an embankment overshadowing Pant Glas Farm (a relic of days when the valley had been green) and eventually, during the First World War, extending up beyond the Glamorgan Canal and the railway that passed behind Maes-y-Bryn, high onto the weeping slopes of Mynydd Merthyr. There, as the decades passed, a chain of brooding pyramids grew to dominate the village. All this because of one man.

Coal entrepreneur John Nixon (1815–99) was the son of a tenant farmer who left school at fourteen and after a little farm work became apprenticed to a mining engineer in Durham. In 1840, so he told the *Merthyr Guardian* in 1860, he had boarded a penny boat from the North of England to London and had been puzzled to see no smoke coming from its stack, despite the strenuous efforts of the stokers coaling the boilers below. He paid a stoker one shilling to let him look at this miraculous fuel, and another shilling to be allowed to feed the fire himself. When he saw with his own eyes how the wonderful substance burnt so cleanly, he asked where they got it. The stoker answered, Nixon told the reporter: 'Meartheer in Wales'. (He could not say 'Merthyr'.) Nixon added, 'I at once made it my business to come down here.'

After first working in partnership with another coal owner, Thomas Powell, Nixon eventually made enough money to begin sinking his own pits. His last and greatest venture was the Nixon Navigation Colliery at Merthyr Vale. The River Taff was diverted to provide more room for colliery infrastructure, and sinking the first of Merthyr Vale's two pits began. At this time, no village called Aberfan existed. This was the industrial frontier: the road, tram and railway on the eastern side of the Taff were in place, but to the west, maps reveal the farms Aberfan Fach and Aberfan Fawr, a few scattered buildings along what is now Aberfan Road (mostly chapels, including the old Smyrna, and pubs) and the chapel of rest in the virgin Bryntaff Cemetery. Thus, the three essential services of any Victorian mining township were in place. For this life, work and drink. For the next, hope and somewhere to be buried.

From commencement of the southerly Pit No. 1 at 2 p.m., 20 August 1869 to the first striking of coal was six long years. Fortunately, the sinker's log still exists. It is a litany of engineering problems, drama (mostly tragedy, but a little comedy) and above all, delay. On the very first day of sinking a mere three feet from surface, water began to run into the pit – not surprising, perhaps, on a river-bank. The first injury (not fatal) occurred four days later. Water and loose sediment were always the main difficulty, with a constant danger of running sand and gravel pouring in and flooding the pit and caus-ing subsidence around the pit mouth. 'Enginehouse, boilers and stack moving', the sinker calmly notes on 5 December 1869. Sinkers were also regularly laid off while engineering problems were put right, and on 6 March 1870 the workers struck in protest when their wages were reduced by ten per cent – to no avail. The sinker's entry, seven days later, reads: 'Sinkers . . . commenced work at a reduced rate of ten per cent.'

Once solid rock was gained, explosive shots came into use, multi-plying the accident potential greatly. Drunkenness was also a problem, the entry for 25 November 1872 reading: 'three o'clock before they got into the pit, through Joseph Randel leaving his work and going up to the new public [house] and getting tite'. Nor did problems always stay fixed. Shuttering, installed to keep back water and loose sediment at the top of the shaft, blew out on 21 June 1873, knocking one William Morris into the pit, where he drowned in a turbid maelstrom of muddy water forty-five yards deep. The sinker reports that he had to grapple for the body with a plumb line, even-tually pulling Morris's body out at midnight, eight hours later. All the men had 24 June off to attend his funeral.

The blasting and digging continued, punctuated by vicissitudes major and minor: deaths, funerals, layoffs, strikes, drunkenness, score settling, short-handedness and petty larceny (John Matthews, caught stealing candles – twenty-one days in jail) until the Four Feet Seam was hit. It was New Year's Day 1875. The arduous process that would eventually give birth to one of the UK's safest and most productive coal mines, and which would continue as such until its final closure on 25 August 1989, had finally hit pay dirt.

First coal was raised on 4 December 1875. Sinking had cost the consortium £255,946 – the equivalent of an infrastructure project today costing about £300 million. When John Nixon retired in 1894, his various coal-mining interests were producing one and a quarter million tons a year; and when he died in 1899, he left an estate valued (rather precisely) at £1,155,069 17s 6d. In terms of personal economic influence within the UK economy, this makes him almost a modern-day billionaire.

The Nixon Navigation's Pit No. 1, which was also the downcast ventilation shaft, was sixteen feet in diameter, with a carriage of just one deck that bore two trams per wind. This was intended for working the Four, Six, Nine and Seven Feet Seams, the winding level being at the Four Feet. Eighty-four yards away, Pit No. 2, up which the mine's stale air was drawn by the ventilating engine, was fifteen feet across. It had a double-deck carriage holding one tram per deck and mainly worked the Nine Foot Seam. The mine's air supply was powered by Nixon's patented reciprocating air pump driving a gigantic centrifugal 'waddle fan' forty feet in diameter, made in Llanelli and capable of pulling 400,000 cubic feet of air per minute through the workings.

By July 1876, the mine was producing 500 tons of coal a day. By 1880, methane gas from the coal beds was being piped to surface to fire boilers and light the offices and stockyard. By 1908, Pit No. 1 employed 1,546 men underground and Pit No. 2, 1,315. Some 526 more worked at surface. In 1913 the pit in total employed 3,575, making it by far the largest colliery in South Wales. Peak employment, 3,685, was reached two years after William Bowen died, in 1924.

In 1929, Nixon's Navigation Coal Co. was taken over by Sir David Rees Llewellyn. Under the new company, Llewellyn (Nixon) Ltd, Pit No. 1 was deepened to 542 yards and Pit No. 2 to 538 yards. With this new investment, in 1934 Merthyr Vale helped the other five in the group to produce 2,100,000 tons of coal. The following year, the mine passed to Powell Duffryn Associated Collieries Ltd, the largest coal-mining company in Britain, formed of many mergers and takeovers. Capitalized at £17 million at its formation, it controlled 44 pits and produced 12,372,000 tons of coal per year – 32 per cent of the entire output of the South Wales Coalfield. Pit No. 1 at that time is recorded as employing 190 surface men and 1,360 underground, while Pit No. 2 employed 100 above and 570 below. Together they were producing 800,000 tons of coal every year – nearly three times the productivity of the average pit in the company.

Powell Duffryn soon became the world's most powerful coal-mining company until Vesting Day, 1 January 1947, two years after Clement Attlee's post-war Labour Government was elected. It had introduced a Coal Industry Nationalization Bill to Parliament shortly after taking office, and saw it pass into law in 1946. Powell Duffryn Collieries received £16 million in compensation – a good deal, because the book value of the company's assets by then stood at only £12.6 million. (It proved especially good for Powell Duffryn's owners, because shareholders only saw £11.6 million.)

According to coal-industry historian and former mineworker Ray Lawrence, the move to state control was also a boon for Powell Duffryn's former managers, who walked into new jobs in the nation-alized enterprise and ended up running the entire coalfield. The National Coal Board, responsible to the people of Britain through Parliament, became the single biggest labour employer in the country, and 110,000 of its workers lived in South Wales. Although King Coal had just, in fact, been made even more powerful by nationalization, fears that board officials might carry on in the high-handed way customary among the old coal owners were tempered by the hope that now, actions would be restrained by the controlling hand of the people, exerted through their representatives in Parliament. That belief proved to be woefully naive.

By 1947, Merthyr Vale Colliery found itself in the NCB's South Western Division, No. 4 Area, Group No. 4. By then, 217 men were working at surface and 843 below ground. Production, from four seams, hovered through the 1950s at around a quarter of a million tons per year. By 1956, a century after my great-grandfather was born and the year that I came into the world, it was still the largest producing pit in its area. In the mine's entire life, from the first day of sinking to the day it was finally closed in 1989, Aberfan's Merthyr Vale Colliery never suffered a single major mine disaster.

This is not to say that workers did not die, and the death toll reveals the wide range of different trades employed at a major mine as well as, through the range of ages at death, the potential length (if spared) of a late Victorian working life. First to die was Thomas Metcalf, a 30-year-old sinker, on 7 September 1871. Five died, in all, during the sinking. Some fell down the pit; others were killed by falling objects or accidents during shot firing. Roof falls were subsequently the most common cause of death, followed by being crushed by trams in the gloom, where a momentary misjudgement of distance could easily prove fatal. On the last day of December 1885, door opener John Jones (whose job, aged 12, was to open the brattices to allow trams to pass – a common first job) was crushed by one of those trams. It had only been just over six months since his colleague David Rees, also 12, went the same way. At the other end of the age scale, Labourer David John was run over in 1886 at the age of 70 – while John Davies, pumpsman, met the same fate in 1893 aged 71. They were venerable men indeed, at a time when the average age of a mineworker at death was about 45. Oldest to die during these two decades of the late nineteenth century was engineman John Evans, who at 76 simply fell downstairs.

And so on, through the ranks of labourers, rippers, shacklers, collierboys, hauliers, roadmen, screenmen, wastemen, carpenters and prop-takers, deaths mounted at the rate of about five or six every year – almost without discrimination as to age or trade. The wagon examiner was crushed by his wagons, and the ropemaker felled by his rope – like Lewis Morgan, thirty-five, who on 19 March 1890 got caught in a sheave wheel at the top of the pithead. The laconic descriptions of these deaths in the annals of Her Majesty's Inspectorate of Mines belie their horror. There are only two ways to be 'caught in a sheave wheel': either between the wheel and the rope, or in the spokes. Both would slice the victim in two.

All Victorian people lived surrounded by, and immersed in, death. The Queen's sombre reign may have had something to do with it, or fed off it; but surely the causes are to be found in the simple statistics. Infant mortality peaked in the late nineteenth century at 150 per 1,000 births; and if poor sanitation was not a problem, the lack of an effective cure for infection could alone ensure that sudden death – from what we would regard as trivial causes – was commonplace. Combine such background statistics with the effects of living in a community with only one major and very dangerous employer and it is fair to say that no Victorians were more acquainted with grief than those who lived close to a pithead, even a 'safe' one. And yet, to its great credit, the Merthyr Vale's hooter never sounded to herald the recovery – or not – of tens or even hundreds of men killed in an instant by an explosion of firedamp or dust, or by the collapse of a packed cage down the shaft. By the standards of its day, the Nixon Navigation was indeed a safe pit, albeit with a death every two months.

I find it hard to imagine what it must have been like to work in a place where colleagues died at the sort of frequency with which today I add my name to staff birthday cards. I have always found a special poignancy attaches to those who give their lives simply in pursuit of their trade. There is real pathos in it, especially if it all happens decorously, concealed from view. The least we can do, it seems, is dub such men 'heroes'. Yet there was to come a day in British history when the words 'heroism' and 'mining' – much spoken in the same breath by politicians of all persuasions until then – suddenly rang hollow. That day was 21 October 1966 – the day of the worst mining-related disaster in British history.

The Aberfan Disaster deserves the superlative for more than reasons of arithmetic. Mineworkers have always accepted the risks they run. The reason that Aberfan was so devastating was the betrayal, a term that has been much used in the context of Aberfan, its dead and its survivors, and not without reason. However, I am speaking here of another betrayal – the sundering of the bonds of obligation that are, or should be, the cement uniting us with forebears and successors alike.

By 1966, the slopes of Mynydd Merthyr, built of the Pennant Sandstone with its shale partings and coal seams mantled in boulder clay from the melting of the glaciers 10,000 years before, supported seven huge spoil heaps containing an estimated 2.66 million cubic yards of mine waste. Highest on the horizon stood the two conical tips, numbers four and five, begun in 1933 and 1945. These symmetrical cones were created with a device called a Maclean Tippler, which spread waste equally on all sides of a central steel tower. The other tips, lower down the hillside, were numbered one, two, three, six and seven, and built out in southward arcs, like rococo festoons, each leading away from the mineral line which brought the journeys of waste-filled drams to their heads.

Such linear tips were created by a crane, running on its own short length of horizontal track along the tip's contour-hugging crest. The crane would pick the wagons up, invert them to spread their contents downslope and then return them upright to a steel plate which guided the empty dram back onto the rails. The ill-fated Tip No. 7 was the only active one at the time of the disaster. It had been started in 1958. By 1966 it contained 297,000 cubic yards of waste, 30,000 of which were of a fine slurry called 'tailings', waste products of the coal preparation plant. The other tips contained no such fine tailings. Tip No. 7 did not need these tailings to make it fail; but their presence did affect the quality of the material that over-whelmed the village.

Even more so than the Pennant Hills on which I grew up, the sand-stones of Mynydd Merthyr are riddled with impervious layers and coals. Spring lines are therefore even more common, as rainwater (almost two metres of which falls every year) percolates down through the sandstone's joints and forces its way up to the surface along the partings. Springs burst out wherever one of these impervious layers is uncovered by the mantle of thick boulder clay. The geological survey sheet of the area clearly illustrates where these layers crop out. The Ordnance Survey base maps, made and remade from 1874 through to 1900 and 1919, and just like those of Carnglas and Llwyn Mawr, are dotted with the marks 'Rises' and 'Spring'. They had been there for 10,000 years, running weakly or strongly depending on the season, but ceasing only in exceptionally dry summers. Despite what was said during the aftermath, everyone knew about them.

The tips of Merthyr Mountain themselves, had anyone taken the trouble to read the signs, were already speaking eloquently of their location's unsuitability and the inadequacy of their civil-engineering design. In fact, in the planning and siting of all seven tips, civil engi-neering was notable mainly for its almost complete absence. Tip No. 4, for example, begun in 1933 and second highest after Tip No. 5, was intended to be a conical tip. No surveys of its eventual design foot-print were carried out, yet from the positioning of its central steel tower it was inevitable that it would grow to cover the source of a major stream. No drainage was attempted, and no effort made to cul-vert the stream courses that would eventually be engulfed. Instead, the stream spring was simply overwhelmed, and so continued to bleed its waters into the base of the growing heap.

On 27 October 1944, after some worrying precursor movements, the inevitable happened and a large portion of the tip failed and slid down the mountainside for almost 600 metres, stopping just 100 metres or so short of the disused Glamorgan canal. If one definition of insanity is, as Albert Einstein suggested, doing the same thing over and over and expecting a different result, tipping practices above Aberfan were truly insane. The 1944 slide, as the tribunal put it, 'pro-vided a constant and vivid reminder (if any were needed) that tips built on slopes can and do slip, and having once started . . . travel long distances'.

The 1944 failure rendered further use of Tip No. 4 impossible, so Tip No. 5, which was to grow to be the tallest, was begun according to the same plan. Very few people, and none at NCB headquarters, ever heard about the slip of 1944. This might sound remarkable, but there had simply been no need to report it, according to NCB regu-lations. The slip had not injured any person working at the mine. As Sir Herbert Edmund Davies acidly observed in the tribunal report, based on that criterion the colliery would not have been obliged to report the Aberfan Disaster either.

Tip No. 5, which became the sole working tip in 1945, was com-menced after no ground investigation or preparation of any kind. Unsurprisingly, therefore, as it grew it too covered a spring and its watercourse downstream to a distance of about 300 metres. By 1951, aerial photographs clearly demonstrate that a large and ominous bulge had appeared on its south-east flank. No precautions against its possible failure were taken during the five remaining years of the tip's life, which ended in 1956. It has never been clear why Tip No. 5 was closed, though the decision was probably made as a result of three factors: incipient slippage, enormous size (706,000 cubic yards) and the fact that it was now partially alight. This burning was what gave Tip No. 5 its ash-white streaks, visible clearly in my father's photograph of 1968.

In 1956 dumping switched to Tip No. 6, a linear tip lower down the mountain, the only one sited north of the mineral line, which built out at right angles to the slope rather than parallel with its contours. No plans were produced, and no precautions against slippage were taken. Tip No. 6 had a short but uneventful life, brought to an end not by failure but by a letter sent by a neighbouring farmer pointing out that the Coal Board was tipping on land that didn't belong to it. A survey was hurriedly ordered to determine what land exactly the board did own on Mynydd Merthyr. As group planning engineer Mr Warwick James Strong soon found out, farmers are rarely mistaken in such matters. The colliery needed a seventh tip.

NCB managers turned their attention once more to land south of the mineral line, immediately downslope from Tip No. 4 – the one that had slipped back in 1944. From its inception, anyone could have seen that Tip No. 7's growth trajectory would eventually take it directly across the slipped material that had come down the hillside from Tip No. 4 – indeed, across the very same watercourse that had caused that earlier failure. But, as the tribunal report said: 'No one gave any thought to the ultimate maximum area of Tip No. 7.' The site was selected by Ronald Neal Lewis, group manager, and Joseph Baker, group mechanical engineer. Neither man had any background or education in tip design, civil engineering or geology. No survey was taken, no consideration of geological or geographical features was given and no guidance as to the ultimate extent of the tip was ever issued to Mr Robert Vivian Thomas, the colliery engineer whose job it would be to supervise it. One of the most telling pieces of evidence ever laid before the tribunal, during its unprecedented duration of seventy-six days, was that on the fateful day they climbed Mynydd Merthyr to approve the siting of Tip No. 7, Messrs Lewis and Baker took no map with them of any kind.

The twenty-first of October 1966 dawned calm and windless. Mist hung low in the valley, hiding the mountain tops from view as people went about their business and their children made their way to school – juniors for the bell at nine o'clock, seniors for half past. The slopes of Mynydd Merthyr stood high above the mist. The first of the men working Tip No. 7 arrived just before half past seven, though without their charge hand, Leslie Davies, who was at the colliery giving his regular Friday report to colliery engineer Vivian Thomas. Mr Gwyn Brown, who operated the crane, and slinger David Jones, walked to the point of Tip No. 7, and saw the pitheads and the boiler-house chimney of Merthyr Vale Colliery poking out through a blanket of white mist.

But when they inspected the active point of the tip, as they did each day before moving the crane up from its parked position ready for work, a more disturbing sight greeted them. The two pairs of rails that yesterday had carried the crane's full weight had today fallen into a pit three metres deep. Not liking the look of what he saw, Brown suggested to Jones that he contact the charge hand down at the colliery. This meant a return journey on foot: because its wires had been repeatedly stolen, management had removed the telephone. Jones set off, leaving Brown, aided by the other gangers as they arrived, to retrieve the tram landing-plate from the hole and move the crane further back from the edge.

Jones found his charge hand, who reported the news to the colliery engineer. An oxyacetylene cutter team was despatched to cut off the overhanging tracks; Thomas ordered Davies to stop tipping and told him that, on the Monday following, he would come to the site himself to find a new tipping place. Davies, Jones and two men with the cutting equipment climbed back up the mountain, arriving at nine o'clock. While they had been away the depression had doubled in depth.

Davies told Brown that they would have to move the crane still further back; but before doing anything, they all retired to the cabin for a well-deserved cup of tea. Brown alone remained. As he stared down from the edge of the depression, he suddenly saw it begin to rise back up. 'It started slowly at first,' he told the tribunal: 'I thought I was seeing things. Then it rose up after pretty fast, at a tremendous speed. Then it sort of came up out of the depression and turned itself into a wave – that is the only way I can describe it – down towards the mountain ... towards Aberfan village ... into the mist.'

His shouts of alarm brought Leslie Davies and the others out of the cabin. He and his mates then ran, with the deafening roar behind them, the sound of trees cracking and a tram passing. In the mist they shouted to each other as they descended Tips Nos 3, 2 and 1: 'All I could see was waves of muck, slush and water ... I couldn't see – nobody could.'

The mountainside farmhouse and cottages at Hafod Tanglwys Uchaf lay directly in the path of the slide and were wiped off the map, killing everyone within. One hundred and forty thousand cubic yards of black slurry then hit the disused canal, fracturing the water main that had been laid along it and leaping the old railway embankment. Now the slip was in the village, where it destroyed eighteen houses, Pant Glas Junior School and part of the neighbouring County Secondary, before finally coming to rest on the Aberfan Road. By that time it was 9.15 a.m. The last child brought alive from the filthy morass emerged at 11 a.m. Bodies continued to be found days later. In total, 144 lives were lost, 116 of them children aged mostly between seven and ten. One hundred and nine perished in the junior school. Of the twenty-eight adults who died, five were, like both my parents, primary-school teachers.

Mr George Williams, a barber in Moy Road – one of the worst affected streets – had been on his way to open his shop. He was expecting a busy day, people wanting to look their best for the week-end. He heard a jet-like roar through the fog but saw nothing until the windows, doors and then the walls of Moy Road houses burst and collapsed 'like dominoes'. Protected from burial by a sheet of corrugated iron, he was rescued later by council workers; and what he remembered most, he told the tribunal, was the hush – 'like turning off the wireless . . . you couldn't hear a bird or a child'. At 9.20 a.m., the hooter at the colliery that had never suffered a major disaster broke its long silence. It had been the last day of term.

Today, responsibility between forebears and descendants is given the grandiose title of 'intergenerational equity' – an incomprehensible term for a simple idea: that present generations should not by their actions jeopardize future ones. It is an expression of the natural instinct of every parent. What happened in 1966 was a mass betrayal of intergenerational equity. For, in slaying 116 children at Pant Glas, it seemed that somehow the sins of the fathers, rather than their blessings, were visited upon the sons and daughters of Aberfan.

The chain of obligation was broken. Parents, who knew that they owed their children a better chance in life, looked upon what had happened and, however undeservedly, felt themselves weighed in the balance and found wanting. Worse, they felt complicit. Nothing makes the knife blade of grief sharper than the alloy of guilt, and as television spread its images across the world, everyone shared the feeling – even those whose complicity amounted to no more than shaking a scuttle of coal onto a household fire. Who indeed could look on as another limp little body was pulled free from the slurry, without feeling that they could and should have done more, however insignificant? Many dropped everything and went – I have friends whose fathers did just that, and who even today cannot speak of it. From all over the world, sent by those who couldn't go themselves, money and toys began to pour in.

Naturally, in Aberfan many people's instinct was to try to ease the pain with which they, however unjustly, burdened themselves, by finding someone else on whom to shift it. Many in the village blamed the gangers who had worked on the tip, men whose conduct was completely and explicitly exonerated at the tribunal. But there was no shortage of more deserving candidates in the tribunal's report – though, as it also admitted, there were 'no villains in this harrowing story ... of bungling ineptitude, by many men charged with tasks for which they were totally unfitted ... decent men, led astray by foolishness or by ignorance, or by both in combination'.

No matter how vivid the condemnation, the people of Aberfan found their pain was not eased. Nothing, truly, could have done that; though it did not help that no prosecutions followed, or that the Coal Board's chairman, Lord Robens, and many others, successfully clung to their jobs. The Aberfan Disaster Fund finally reached £1.7 million, though few who gave would have expected that some of their money – £150,000 – would be used to clear away the remaining tips, after the NCB and the Treasury refused to accept full responsibility for doing so.

Nor did it help that, despite being a clear breach of charity law, Robens's raid on the fund went unchallenged by the Charities Commission – and remained without redress until 1997. There is a myth that, once upon a time, there was an age when men of honour fell on their swords. It may be true, but if such a time ever existed, it was over long before the 1960s. The tribunal report was a masterpiece of judicial writing; but it too was betrayed by the events that followed – or rather failed to follow – its publication in July 1967. Today, many people believe it to have been a whitewash. It was nothing of the kind, but heads failed to roll.

That hardly matters any more. The Aberfan Disaster not only ripped the heart out of one small Welsh village, it sucked the life out of an entire industry. The eerie silence that followed as the landslide came to rest also fell upon the phrase 'the heroism of mining'. This rhetorical chestnut had stubbornly survived its own debasement in generations of political speeches, and still resonated with ordinary people. Now drained of meaning, it passed completely out of use. Even 'the dignity of labour' seemed mere empty words as a generation, in the extremity of its grief, cursed their forefathers, threw down their monuments and turned their faces to the earth in shame.

At the time of writing, two years after the fatal accident at Gleision, a trial date has finally been set for mine manager Malcolm Fyfield and two directors of MNS Mining Ltd, Maria Seage and Gerald Ward. Fyfield faces four counts of gross-negligence manslaughter, while the company faces four counts of corporate manslaughter. The trial will straddle Easter 2014 and is expected to last three months.

4

Headstone

... to pass unheeded into the limbo of forgotten things.

Sir Herbert Edmund Davies, *Report of the Tribunal appointed
to inquire into the Disaster at Aberfan*, para. 48

Anyone who has worked in the UK oil industry, especially through the late 1970s and early 1980s, will have come to know the Granite City. Even if their work did not take them through the airport at Dyce and its strangely temporary-feeling hotel, its corridors full of piped country and western music and stalked by jet-lagged, bow-legged men in ten-gallon hats, they will have encountered the greatest hazard to be faced by any oil worker in his or her risk-filled life – the staff Christmas party in Aberdeen.

In December 1982, having abandoned academe for the mercies of the oil industry, I found myself at the Royal Darroch Hotel in the Aberdonian suburb of Cults. It was a disproportionate, slab-sided 1970s block of brick and concrete, looking rather like an ocean liner beached among the villas, built to profit from the expected increase in traffic as the North Sea boom took hold. I got there very early on the morning of the bash, arriving by sleeper train from King's Cross.

Unable to sleep on the sleeper, I was exhausted, but found that my room would not be ready for another seven hours. What followed was a typically disastrous evening from one's mid-twenties, that ghastly period in a modern graduate's career when, booted from the back door of education, one finds that scaling one dung heap, just a little higher up the mountainside of life, is not greatly preferable to any other.

Dinner that evening was my first formal appearance in the role of Significant Other, and it was less than a roaring success. Depressed by the turn our lives had taken, we consumed industrial quantities of Californian red wine, a substance whose early 1980s novelty had yet to wear off, and at some juncture, she departed to bed – each of us wondering how, despite jobs and money, life had suddenly become so miserable. We had little enough right to complain. Folks of previous generations, and many of our own, would have killed for our problems. But the education that others craved had infantilized its ungrateful recipients. And as though to rub it in, ten months after our visit the Royal Darroch Hotel was utterly destroyed in a disastrous gas explosion and fire that killed six guests and staff.

As fiasco degenerated into disco, I decided quite suddenly to go for a walk. It was past midnight, with a good bright moon. Braced by the chill, I set off – a fine sight, no doubt, in my velvet dinner jacket – and headed eastwards along the North Deeside Road. I walked for a couple of miles, having no particular clue where I was heading, tipsily confident of finding my way back by the the simple expedient of repeatedly turning left.

I had made one left turn already by the time I found myself at a T-junction where, having walked briskly up the hill, I paused to get my breath. Over the road I saw there was a wooded embankment, right behind a low retaining wall and dilapidated fence. An area of waste ground, I assumed; probably some former industrial site, screened off by the tree-planted bank. I was not wrong; though the maturity of those trees should have told me something. I was naturally drawn towards it; it looked like my kind of place, and I felt I had to know what lay beyond. I crossed the road, surmounted the low wall and found my way through the poor torn fence into a thicket, beyond which I climbed into more open wood.

Halfway up the bank, which memory has made much higher and steeper than it proved when I revisited it recently, something strange and unnerving happened. It was already a frosty night, but I suddenly felt an icy waft as cold as if it had come from the Ice Age itself. My breath condensed even more voluminously on the air as the sudden rush of deep cold cascaded over the bank and through the trees, stirring the leaves. I waited as it passed; and then I noticed something in the quality of the silence – a presence, vast and hollow, over the bank. I grabbed hold of a sapling and climbed to the crest, where I found myself staring down a more or less vertical drop into the biggest, blackest abyss that I had ever seen.

The apparently sheer rock walls (which actually taper inwards slightly) plunged straight to the utterly smooth waters of a lake far below, whose surface, black as sump oil, reflected nothing at all. Moonlight shone from the leaves of trees clothing the overhangs and the grey granite below them, but the water sucked all light from the air. A deep silence hung over the chasm, broken only by invisible night birds that now and then chip-chipped in the canopy.

I understood immediately what had just happened. A passing breeze, blowing towards the sea, had disturbed the surface of this deep pond of icy air, which had settled all the raw day in that sunless hollow and slopped just a little of that brimming, dense cold over the bank and sent it rolling downslope. I stared for what seemed a long time at the bottomless black (the experience fitting my mood well), wondering how a thing like this, resembling some great volcanic vent, should come to be lurking lethally just over a flimsy fence in the Aberdonian burbs, the neat granite villas and bungalows with their trim little gardens pretending, so innocently, to know nothing.

I was gripped by a horror – not, I like to think, inferior to that which seized the poet in his elfin pinnace – took fright and lurched back down to the safety of the pavement, trying to recall how many more left turns I should make to get back to the Royal Darroch. As with Mr Wordsworth, my sight of this huge and mighty form troubled my dreams for many days. I found out later that what I had stumbled upon, frequently hailed as 'Europe's deepest hole' and reputedly ninth deepest in the world, was Rubislaw Granite Quarry.

Until about 1740, Rubislaw had been a small hillock a few miles outside the city. A stonemason from Montrose called David Barrie was engaged to ascertain the potential for working the good granite below the weathered rubbish on top, known as 'barr'. As a result, the council took the land but eventually, and short-sightedly, sold it in 1788. Bought by the ancient family Skene of Rubislaw, it was leased to John Gibb, an engineer who had been brought in by engineer Thomas Telford to superintend his Aberdeen harbour works. Gibb and his heirs held the lease until 1889, by which time Rubislaw had supplied rock for the docks of Southampton, Portsmouth and Sheerness, and most famously, Robert Stevenson and John Rennie's lonely Bell Rock Lighthouse of 1810. Standing eleven miles offshore from the Firth of Tay, it is the oldest sea-washed lighthouse in the world.

By the time the quarry closed – one of its last big jobs being to supply cladding for the base of the NatWest Tower in the City of London – it had reached a depth of nearly one hundred and fifty metres. Sunlight never reached its floor, which lay sixty metres below sea level, where the cold and dank prevailed even on the hottest day.

The working face was then nearly twenty metres high, and wept continuously, flooding being prevented by half an hour's pumping every morning (and several hours on a Monday).

Aberdeen's 470-million-year-old granite mass covers a wide area, and was exploited in many more quarries besides Rubislaw. In the 1930s there were working pits at Kemnay, Toms Forrest, Sclattie, Persley, Cluny, Backhill, Balmedie and Kincorth. But Rubislaw was worked for longest, producing some six million tonnes during its 230-year life. Business came to an end just over a decade before I first saw it, not long after supplying the second phase of the NatWest Tower job. The closure process began in May 1970, and the quarry has lain abandoned ever since. Plans were developed since to turn it into a museum, a climbing centre, a rubbish dump, even an underground city or a nuclear bunker. But nature had her own idea. Abhorring the vacuum, she slowly filled it up with millions and millions of litres of pure, clear water.

In my ludicrous velvet dinner jacket, I had been staring at Aberdeen's old source of mineral wealth from the perspective of someone working in its new – the industry that had come along just in time, just as quarrying and fishing were dying, and shocking the thrifty folk of Aberdeen with its splashy cash-rich ways. Ever since that night, and because I know that Aberdeen dominated the UK's granite industry for decades, I have held on to a pleasant though unprovable theory that just two of Rubislaw's six million tonnes of quarried grey granite eventually found their way onto William Bowen's tomb, and that on that frosty night, the obelisk, capstone and shrouded urn had been accidentally run to earth by his descendant.

As the notices pinned to the wooden posts surrounding William's dismantled tomb had urged, I had telephoned the Cemeteries Office of Merthyr Tydfil Borough Council. I had been momentarily confused to discover that in the meantime this office, which I found to be a model of efficiency and consideration, had through no fault of its own been rebranded as 'Bereavement Services'. Be that as it may, the employee of the former Cemeteries Office explained that the council had conducted safety surveys of all its cemetery holdings over a period of years. Many contained memorials dating from the late nineteenth and early twentieth centuries that were in a dangerous or potentially dangerous condition, especially if they were no longer being maintained. I explained that I was interested to restore and maintain my relatives' tomb. 'In which cemetery?'

I was able to say not only which cemetery, but give the precise plot number, though this did not surprise the officer as much as I had hoped. I explained that, to the best of my knowledge and barring miracles, I was the sole living relative of the deceased, including all subsequent occupants (my great-aunt Catherine and her husband John Prosser Lewis). These were all the burials marked by memorials, though later I discovered that John and Catherine had had two sons, both of whom had died tragically young but, perhaps mercifully, unmarried. They too were buried in the plot, unmarked and in earth rather than brick, during the lean years of the late 1920s and early 1930s. No, I was told, there was no requirement for me to prove my relationship. No, nor to send a copy of my great-grandfather's receipt. However, I should take care to ensure that any and all work be carried out by a mason on the BRAMM. They would send me a list.

I had never heard of BRAMM, but years spent working in education and journalism had made me good at guessing the meanings of unimaginative acronyms, and I had a shrewd idea that this one stood for something beginning with 'British' and ending in 'Monumental Masons'. It was in fact the British Register of Accredited Memorial Masons, whose members alone are permitted to work in cemeteries belonging to Merthyr Tydfil Borough Council.

The tomb's original maker had been a Mr Pryce, whose name I

found modestly inscribed on a good many Aberfan burials dating from the early twentieth century. Mr Pryce himself doubtless lay in the eternal shade of one of his own creations by now; but maybe his firm was still going, passed on from father to son. I searched the BRAMM list for it in vain. My second thought was to keep it local, and choose the mason closest to the cemetery gates. If I was going to spend some money, it seemed only right that I should spend it in the community of which William had been such a pillar.

After inspecting the wreckage, the local guy put his ears back and refused the fence. 'Big tomb, that,' he said. 'That'll need one of the big boys, with cranes and whatnot. Better talk to Mossfords in Cardiff.' Mossfords knew the site, had the capacity to deal with large blocks of rock and were in fact the sculptors responsible for the elegant arcades of shining white marble that now mark the graves of the children who died in the 1966 disaster. (They were also, I later discovered, founded in 1821, earlier even than the late Mr Pryce.)

The local guy was not mistaken about the scale of the job. Before I had even started to scope the work properly, it was clear that the whole tomb would have to be dismantled to ground level and completely rebuilt. I knew there was a brick vault beneath – it said so on the receipt – and I knew enough civil engineering to suspect that this would first need to be inspected, and if found to be in satisfactory condition, a reinforced concrete ground-plate would be laid on top. Before any of that could begin, though, all the stones would have to be individually cleaned and repaired, presumably in workshops off site, and the inscriptions re-enamelled. Then, labelled carefully like the former London Bridge (which was, incidentally, built of Rubislaw granite) or the Temples of Ramesses II at Abu Simbel, the whole thing would be reassembled, bit by numbered bit. To ensure that no vandals came to any harm, all the elements would be fixed firmly together – which would, I imagined, mean drilling them top and

Dismantling the tomb, 2012. After the removal of the pedestal in grey granite,
workmen open up the box tomb (Forest of Dean Stone). The field in the
valley below is the site of the former Merthyr Vale Colliery, Aberfan.

Reassembling the tomb, 2012. Artur (seated) and Wojtek (Mossfords Ltd)
at work on restoration after the component parts had been cleaned
and repaired at workshops in Cardiff.

bottom and inserting some sort of dowel or pin, presumably in stainless steel. These assumptions proved to be well founded.

This would mean more than just 'cranes and whatnot'. It meant trucks, brawny-backed men, skilled craftsmen and weeks of labour. I began to lament this obsession (of past generations, but also my own) with memorialization; but I felt that I had to carry out these repairs, and not only for William's sake. I had to do it for the sake of my paternal great-great-grandfather, Edward Nield of Mossley, Lancashire (1833–1905), who was an ecclesiastical stonemason, and at least one other, less direct paternal ancestor who worked on memorials.

The great Victorian age of building dominates our cities of the dead every bit as much as it dominates so many British cityscapes, especially cradles of industry and commerce such as Manchester. It is also especially true in places like Aberfan, which did not exist as settlements before the 1880s, and where Victorian tomb designs clung on for decades after the old Queen died, lagging conservatively behind the cutting edge even of an art form whose fashions move as slowly as the great rock masses in which they take shape. Victorians, after all, went on dying for decades, and took their tastes with them to the grave.

Memorial markers chosen by Victorian Nonconformists like William Bowen tended not to bear the traditional cross, or weeping angel in white marble. They embraced instead the stark and monumental. Roughly hewn rock masses surmounted by an open book symbolized simple faith. A broken classical column might indicate a young life cut short, or perhaps the futility of this world's works when compared to the eternity into which we go. Egyptian obelisks emphasized timelessness. And although he died in 1922, William Bowen was a solid Victorian; and, true to himself, he chose a traditional design for his memorial that had emerged almost unchanged from the nineteenth entury – a style known as the 'pedestal'.

All pedestal tombs consist of three basic elements: a plinth, a shaft and, in most cases, a 'termination' – some device or piece of statuary (usually, as in William's case, a funerary urn). Most, especially later in the nineteenth century, were carved in granite. A typical design would consist of a short ziggurat of three or four square plinths of diminishing size (which may or may not bear inscriptions) followed by a square-sectioned, tapered obelisk carrying the main dedication. This would be topped by a 'cap mould' – also square-sectioned like the obelisk, and 'decorate' (as a mason would describe it, meaning relieved by fancy, etched designs). This would usually support an urn, with or without a drapery shroud, to heighten the grief.

Although standing over two metres high on its own, and weighing many tonnes, this monument was mounted upon a 'box tomb' – a stone chest, designed, like all box tombs, to look as though it contained mortal remains. Actually, it served two quite different purposes – the practical one, in steeply sloping Bryntaff Cemetery, of providing a level surface for the granite elements to stand on, and also the theatrical one of lending even greater height, at lower cost, to increase the tomb's presence.

This stone chest, fashioned in buff sandstone, was set in the middle of the square triple plot, at right angles to the drive. The plot was enclosed by kerbstones in the same sandstone, while to the front, below the kerb, a small retaining wall of sandstone shoddies two courses deep brought the monument to the feet of the contemplative visitor. The sandstone did not look local to me, but there was something familiar about it. Fine of grain, and cut to a smooth ashlar by Mr Pryce, I had seen nothing like it in the South Wales Coalfield. But as Simon Morgan – managing director of Mossfords – confirmed, it was after all a form of Pennant Sandstone, one still quarried today for building and sculpting and known as 'Forest of Dean Stone'.

The Forest of Dean, lying on the edges of Wales just west of

Gloucester and bounded by the Rivers Severn and Wye, is a unique area with a distinctive history closely connected with mining, smelting and quarrying. Like the much larger South Wales Coalfield next door, it is a structural basin where coal-measure rocks lie enclosed within a pie dish of Carboniferous limestone. The combination of iron ore and abundant wood gave rise initially to a smelting industry, which from about 1818 switched from wood to coal as its main energy source. And so it was that, here as elsewhere, iron-ore mining and smelting gave birth to coal mining as new technology made smelting with coal possible.

Because of the importance of ground conditions and earthworks to the outcome of battles, warfare and geology have frequently advanced hand in hand. In one of the oldest manifestations of this uncomfortable fact, the skills of the miner too have often been called upon. Miners were brought in to dig tunnels in the chalk beneath the Western Front during the First World War, creating miles of fighting tunnels in such places as the 'Glory Hole' on the Somme, and in order to lay explosives beneath enemy lines. In 1296, when England invaded Scotland and so began the first War of Scottish Independence, iron mining was already well established in the Forest of Dean. Miners living in the Hundred of St Briavels (the forest's administrative capital, established by the Normans) were called up by Edward the First to help him break the siege of Berwick-upon-Tweed by undermining the town's defences. Successful action by them in 1296, 1305 and 1315 earned the king's gratitude, and he granted to true-born foresters and their descendants the rights to free mining there, in perpetuity.

This ancient legislation was later enshrined in a special Act of Parliament called the Dean Forest (Mines) Act 1838, stating that: 'All male persons born or hereafter to be born and abiding within the said Hundred of St Briavels, of the age of twenty-one years and upwards, who shall have worked a year and a day in a coal or iron mine within the said Hundred of St Briavels, shall be deemed and taken to be Free Miners.' This law, amended since 2010 to purge it of sexism, remains in force today (the 1946 Nationalization Act specifically excluded the forest). Since maternity facilities at Dilke Hospital were closed, it has become rather unusual for people to be born within St Briavels Hundred, though a handful of mines still operate, many diversifying into tourism and 'colour mining' – the production of natural ochres – which has also long been a feature of the forest's mining culture.

If a free miner wishes to open a stone quarry in the forest, the procedure is the same as if he wished to sink a mine. First, he (or, now, she) must apply for the right to take stone from a specified area – known in the forest as a 'gale' – at the Gaveller's Office at Coleford, where all such business is still transacted. Assuming there are no objections, the quarry is leased (renewably) for twenty-one years, subject, usually, to a number of planning and other conditions, such as not impeding footpaths or impinging on another's land, and, of course, paying a fee to the Crown. Today, several quarrying firms operate in the forest, producing stone for paving, sculpture, structural and monumental uses. Like the Pennant Sandstone elsewhere, Forest of Dean Stone comes in several colours: buff, lilac, green and blue, depending on the oxidation states of iron in the different minerals coating and cementing the sand grains together; but what strikes you immediately is its wonderfully even texture.

This is Pennant Sandstone scrubbed up, in collar and Sunday suit, on its best behaviour. Free of any coaly flecks, this freestone has no pronounced weaknesses in any direction. A report from the October 1896 edition of the journal *The Quarry* tells of a single, intact block of Forest Blue at Knockley Quarry that measured 7 feet 6 inches (2.3 metres) thick and weighed an estimated 600 tons. Such high-quality stone, while it could never replace inferior, but locally won and hence cheaper Pennant for building workers' cottages in the South Wales Coalfield, nevertheless found ready markets, at home and overseas, on prestige projects.

Cardiff Castle used it – William Burges's nineteenth-century pseudo-medieval masterpiece, commissioned by that great coal king, the Marquess of Bute. Burges's famous stone animals, which seem to be escaping over the walls of Bute Park and threatening to invade Duke Street, are carved in it. It was fitting, I thought, that my great-grandfather, who had hacked his way through its uncouth coalfield equivalent all his life, should have his memorial mounted on this tidy-edged version. It was a subtle idealization of a life.

So good was Forest of Dean as an urbane, smooth-sawn building stone, that during the late nineteenth century it was even exported as far afield as India and China ('for the purpose of assisting in the oper-ation of drying tea', according to reports). A hundred years on, that flow of rock has changed direction; but during the height of Empire, British building stones of every kind found their way all over the globe. Imports, on the other hand, were sucked in – largely by the simple incapacity of domestic quarries to satisfy all the demands placed upon them by the frenzy of paving, kerbing, dock-building and memorialization that both expressed and bolstered the civic pride of British Victorian boroughs.

The worldwide transportation of stone is, therefore, a lot older than we might at first expect. It began, in a small way, during the building of the world's great empires from the late eighteenth century, culmi-nating in the Victorian age. Victorian imperialists colonized not only by conquest but by mapping and geologizing, and early surveys soon discovered that suitable stone for building harbours and defences existed aplenty in these distant outposts. But although it was there, it was also of little practical use. The presence of a reserve of Earth

materials is only the first requirement. Without an established industry, it cannot be said to constitute a 'resource'.

In the industrial age, the factor that has always initially dictated the
source of stone for big infrastructure projects like docks and light-
houses was not geology alone, but geology plus infrastructure – the
existence of an established skills base. At first, well-tooled Britain
exported its granites to granite-rich India. Now the situation is
reversed; and quarries abroad, once the stopgaps, supply almost all
our demand.

Tombstones are the ultimate luxury item, their heavy emotional
loading enabling them to command the high price necessary to buy
the time, skill and labour needed to cut, carve, polish and transport
them. Economics have to be very tough indeed before departed loved
ones do without commemoration (as they were when my two mysterious cousins once removed died during the Great Depression and
were ignominiously buried in earth under William Bowen's obelisk).
Nevertheless, until the nineteenth century, apart from the marble
baroque monsters imposed by local nobility upon the transepts of so
many humble English parish churches, most ordinary gravestones still
tended to be made from reasonably local stone, where available. But
not just any old stone will do for a tomb.

The simplest headstone is basically a paving stone set upright. But
even making a flat slab that does not shiver along planes of weakness
already narrows the range of stones one can use. That range is diminished further if you wish to carve detailed lettering or ornament, and
still further by the sterner demands of sculpture. In Ilston Churchyard
in Gower, Francis Kilvert would have seen a number of gravestones
dating from the seventeenth and eighteenth centuries. The grandest of
these is of slate, the nearest sources for which would have been
Pembrokeshire or North Wales, yet which would still only have fallen
within the means of the wealthiest local farmers and landowners.

Joseph Pryce of Gelli-Hir was one such, and his memorial stands fixed to the church's south wall, behind the tower, shaded by the great yew and protected by some serious railings. *'Quod mori potuit?'* the stone is headed – here was a man to whom death could do nothing: 'religious without hypocrisy, humble without meanness, charitable without ostentation, humane, hospitable, an affectionate husband, indulgent father, a sincere friend, impartial magistrate, his life exemplary, his death happy'. We are asked to believe that 'Unfeigned piety accompanied him through every hour to the last moment of his life; supported him with assurance of hope, the source of peace, and earnest of a blissful immortality.' It takes a lot of money to afford a stone large and durable enough to bear a posthumous reputation of that order.

During the relaxing time after I had submitted my doctorate, while still waiting for the pall of employment to descend, I set about recording the epitaphs in Ilston churchyard for a local history journal. What struck the geologist in me at the time was the lack of any gravestones in local Carboniferous limestone. There was more than one reason for this, the main one being that the limestone (a large quarry, now a nature reserve, lies just outside the village) was simply not suitable for slab-making in this area. The vice-like stresses that created the great downfold of the South Wales Coalfield also imposed pervasive lines of weakness, called joints, that criss-cross the rock and cause any limestone bed to fall to rubble.

You don't need freestone to make slabs; but any cross-cutting joints in a rock stratum are fatal to its use either as tombstone or flagstone, and tectonized Carboniferous limestone is quite useless for this purpose. This is a shame, since the original 'rock of ages' described in the hymn was an outcrop of Carboniferous limestone in the Mendip Hills. It is said that in 1763 the hymn's author, the Rev. Augustus Montague Toplady, was forced to take shelter during a sudden storm while out on his parochial round and did so within a rocky bluff in Burrington Combe, Somerset (which today bears a plaque commemorating the event).

Not all calcareous rocks – and not all Carboniferous limestones – suffer from this jointing problem. In parts of the country nearer sources of more tractable well-bedded limestone, one finds it used frequently in eighteenth-century tombs, commemorating the deceased of the 'better sort'. It is easy to see why; the limestones are local, relatively easy to carve with hand tools and can lend themselves well to fine lettering, or perhaps a naïf bas-relief. But their biggest disadvantage is that they weather readily, and their careful inscriptions are soon lost. Impermanence is not a welcome quality in gravestones.

Until well into the seventeenth century, only the very wealthy received stone memorials of any kind. For them, money being no object, exotic materials could always be brought to site. In areas where good building stone is scarce, old markers for the graves of long-forgotten folk have often been relocated, used for flagstones, or incorporated into church walls during repairs or rebuilding. For example, the Anglo-Saxon Old Church of St Andrew, Kingsbury, near London, was originally built with no stones that came from further afield than Kent. Yet one relatively exotic rock-type is found there.

A former memorial, of possibly sixteenth-century or even older date, was fashioned in Purbeck Marble, an unusual Upper Jurassic green limestone laid down in freshwater, full of the remains of tiny snails and much used throughout history for ornamental work. It was quarried and worked near Corfe in Dorset, transported by sea to London and then by road to Kingsbury – all to preserve the memory of a local magnate. But at some stage in later history the stone, rubbed clean of its inscription, was repurposed to serve as the church doorstep. How indeed are the mighty fallen; the only true immortality is to be found in descendants, and no stone, no matter how hard, endures for ever.

For most of the last thousand years, burials were not marked in any permanent way. Today, Ilston's graveyard, which has seen more than a thousand years of inhumation, is apparently dominated by burials from the nineteenth century – the very parishioners to whom Sterling Westhorp would have preached. Victorians entered and exited the world in more or less the same numbers, and with similar demographic spread, as their ancestors did in previous centuries. The main difference lay in the fact that they were the first people to have access to sources of concentrated energy, mainly coal. The injection of the stored energy of the past into the present created both the wealth and the physical means to move truly durable rock around the country – even between countries – for the sole purpose of keeping faith with the dead. And of all the stones that transport now brought within almost everyone's reach, none spoke more strongly of eternity than granite.

So I now knew where William's Forest of Dean box tomb came from, and I knew that it was custom-made to fit plot 1/17. The granite elements perched upon it, however, would have been bought off a pattern-book and delivered to the mason ready-made, awaiting only their inscription. Where did they come from? My preferred assumption had always been Aberdeen, but determining the provenance of granites poses a ticklish problem. Granites are coarsely crystalline igneous rocks, rocks that formed 'by fire', solidifying very slowly from the molten state deep beneath mountain ranges in massive, intrusive bodies that may be miles across. These bodies are known to geologists as 'plutons' after the deity in Greek cosmogony who, when Zeus became ruler of heaven and Poseidon the sea, assumed the throne as the terrible, brooding, violent ruler of the underworld, with its riches – and its dead.

I do not suppose that William Bowen paid much attention to Greek myths, being too bound up with Hebraic ones. But I find it satisfying that the stones of his main monument were formed in a pluton, named after the god of the underworld, of wealth and death; the world he travelled through with his Davy lamp all his working life, just as Helios Pluton, the midnight sun god, who (in one version of the cosmogony) passes through the underworld each night towards morning. Helios Pluton was not venerated widely, but in the first or second century AD a temple was dedicated to him in a certain city of Asia Minor . . . called Smyrna.

The immense size and great depth of plutons account for the slowness of their cooling, and that slowness in turn is what gives granites their coarse texture, as each crystal has a long time to grow. Sometimes, granites may have a more complex cooling history, and different phases of crystallization may occur at different rates. Such history gives rise to rocks where some very large crystals appear to float in a finer matrix. These are known as 'porphyry', one well-known example in Britain being the pink granite of Shap Fell, Cumbria, with its large, fleshy, widely separated feldspar crystals, looking like canapé squares of smoked salmon.

To a stonemason, any rock with visible crystals is 'granite'; but this is not what a geologist means by the term. To a geologist, granite is always light in colour, because the melt from which it formed, enriched in certain elements, depleted in others, gave rise to light-coloured and less dense minerals such as feldspar. Feldspar can vary from white to pink; often, as at Rubislaw (and to the consternation of those who would identify them), within one quarry. Among these, one can see clear, glassy quartz and flecks of black mica. Although some other minerals may sometimes be seen in granites, they are mostly light-coloured.

To a geologist, there is no such thing as a 'black granite', a term you nevertheless find in masons' catalogues. Coarsely crystalline igneous rocks that are predominantly dark have geological names of their own – the dark equivalent of granite being 'gabbro'. Geological names go by colour as well as texture, because colour speaks most eloquently of that all-important factor, magma chemistry.

Granite masses mostly form during the process of mountain building; they are the result of what's going on in deeper, hotter and more intense regions, while the same tectonic forces are folding more superficial rocks into structures like the South Wales Coalfield. To be exposed at surface, as they are in and around Aberdeen and Peterhead, tens of kilometres of overburden have to be eroded away. Granites are found chiefly where ancient mountains have been completely worn down, grain by grain – like the inscription on a forgotten tombstone – by the rivers of time.

In Britain, granites have been worked in England, Wales and Scotland, though granite quarries were most common among the older rocks typical of the North and West, where erosion has had the longest time to do its work. Merely having good granite does not mean you have an economic resource: location is all. And so the regions that were always most closely associated with granite quarrying tended to lie close to the coast, allowing the rock, or its carved, shaped and smoothed products, to be exported most cheaply. In Britain this mainly meant either Cornwall or adjacent parts of Devon (where A. L. Rowse's Uncle Joe once 'made the jumper ring', the jumper being a long steel bar used for drilling holes before the advent of powered tools), or the coast between Aberdeen and Peterhead in Scotland's North-East.

Sir James Taggart, Lord Provost of Aberdeen from 1914 to 1919, stone merchant and raconteur, came to the Granite City aged sixteen to be apprenticed as a stone cutter. After spending time in the US, he established his own merchant business in the Great Western Road in

1879. He once famously described the principal exports of Aberdeen as 'fish, granite and brains'. Barely a century later, the granite and fishing industries had withered and the flow of brains reversed, as geologists flocked there from all over the world in search of North Sea oil. But although none of that had been foreseen when Taggart died in 1929 aged eighty, the writing was on the wall for the once all-powerful British granite industry.

My assumption that my great-grandfather's monument had been sourced in Aberdeen took a knock when my monumental masons, in their survey of the tomb, referred to the granite as 'Indian Grey'. The names that stonemasons give to rocks do not only differ from those used by the geologist; they are chosen for commercial reasons. In 1932, for example, imported granites from Finland, described in a consular report, were marketed in the UK under the trade names 'Balmoral Red', 'Birkhall Grey', 'Braemar Grey', and so on, in the no doubt well-founded belief that their Finnish names would not be quite as catchy. So I asked whether the firm really meant that the granite of William's tomb might, as long ago as 1922, have come all the way from the subcontinent, or whether 'Indian Grey' was merely shorthand for granite of that colour and texture, based on the most common modern example. But no – the firm believed that the granite had likely been quarried in British India and exported, ready-worked to order, based on patterns supplied by the UK monument trade – as had been the way of things for decades.

That such a trade exists today is beyond question. 'Black granite' from South India, grey granite from the North, are staples of the monument industry. In fact, as many in that profession have told me, the supply line has grown even longer than from India to the UK. Rough-hewn granite from the subcontinent is now frequently exported to the UK via China, because – despite the extra transport – working the stone there is cheaper than doing it near to where it is quarried. I was surprised, however, by the idea that India could already have been exporting finely worked stone as far back as 1922, so I decided to investigate.

Aberdeen was truly capital of the UK quarry industry, and from the mid-1890s the national trade magazine, the *Stone Trades Journal*, maintained a special correspondent, William Diack, to report every month on the ups and downs of business. Diack's feisty reports demonstrate that the health of the stone trade was largely controlled by three main economic factors – public works, private building and war.

War was always bad for trade at first, removing skilled men to the front and dampening business generally. But wars don't last for ever, and in their wake they bring their own cure in the form of a frenzy of patriotic memorial building. General post-war recoveries in trade likewise helped granite quarries and masons because they led to the commissioning of confident, prestigious commercial buildings, which (at least at pavement level, and much like tombstones) tended to favour the polished solidity of granite, as a token of their vaunted permanence.

Against such volatile influences, public works, demanding mostly crushed rock for roads and roughly hewn kerbstones and setts for cobbled streets, provided relative stability (though they too were not immune from the more serious economic ups and downs). Price is the ultimate driver in a free global market, and as the twentieth century proper got under way after the close of the First World War, the UK granite-quarrying industry found that its economic footings were being progressively nibbled away. Imported stone began to threaten UK producers, not just in their export markets, but on their home ground. Like all trades, quarry producers are quite capable of campaigning at home for the very protectionism they vehemently decry in other countries, and the immediate post-war period in Britain provided a classic example.

While large-scale granite quarrying struggled on until the early 1980s, for most of those sixty years the same questions were asked over and over again about how the industry could be 'saved'. No satisfactory answer was ever forthcoming, because there was none to give. Death had entered the walls of granite quarrying from the moment the Armistice was signed, and for much the same sorts of reasons that many other relicts of British country life that had burbled on, like Hardy's tiny rivulet, through war and peace for a thousand years. Habit and sentiment were not enough. Since a successful granite-quarrying industry is mostly a matter of infrastructure, economically less developed nations with abundant granite were poised to take over. India has no shortage of suitable granites (however defined), and yet for decades, as we have seen, granite from Aberdeen and Cornwall was being shipped out to build harbour walls and breakwaters not only in India and Sri Lanka but also such places as Argentina, Singapore and South Africa.

The ancient world's centre for granite working was Egypt, in the great obelisk quarries of the Upper Nile. The art of granite working, lost after the days of the Ptolemies, was rediscovered in Britain when Egyptian polished stones, millennia old, went on display at the British Museum. The source of these rocks, and the ancient industry that worked them to such perfection, was Nubia – an area now straddling southern Egypt and northern Sudan.

In April 1898, a battle took place between combined British and Egyptian forces and Mahdist Sudanese rebels, supporters of Muhammad Ahmad bin Abd Allah (1845–85), who had proclaimed himself Mahdi, or Islamic messiah. The revolt of his supporters against colonial rule led to the Second Sudan War, or 'Madhi Revolt', as described by Winston Churchill in his book *The River War* (1899). The battle of Atbara, at which the British were victorious, saw the death of many soldiers of the Queen. But to commemorate them, it was not to the nearby granites of Nubia that the Regiment turned – for by then no quarry industry existed there. Instead, Aberdeen grey was shipped out and, in William Diack's words, 'conveyed over miles of trackless African desert and erected on the battlefield in memory of the Scottish soldiers who fell'.

The advantage enjoyed by British quarries by virtue of their long-established infrastructure could not last for ever, and by the early 1930s serious worries were setting in. The post-war boom had bust. And while in 1932 British quarrymen extracted a stupendous 9,213,697 tons of granite from their native soil, that huge figure was down by a million and a half tons on 1931. All areas, from Leicester-shire to Wales, Cornwall and Scotland, showed a decrease. Aberdeen was the only bright spot, its quarries raising production against the national trend (to 402,006 tons from 367,179 the year before, which itself had been an increase on 1930, before the economic downturn had begun to bite).

The reason for Aberdeen's success was that its quarries produced a wide variety of different types of worked stone. This shielded it from the most serious aspect of market downturns, namely government austerity measures, which cut off the industry's bread-and-butter work supplying roadstone, cobbles, setts and kerbstones. While the number of workers laid off in 1933 had doubled over the previous year (to between forty and fifty – hardly enormous), this was a big improvement over 1931, when the number out of work had been over 300.

In spring 1932, having listened to repeated complaints from stone producers about the dangers posed by foreign imports, the Board of Trade introduced a fifteen per cent tariff on imported stone (later reduced to ten per cent on unworked quarry blocks). To monitor the effect of their new import barrier, for the first time the Board began to collect detailed statistics. This enables us to build up a picture of where this foreign competition was coming from, and which sectors of the market were most affected (the statistics distinguishing helpfully between granite for setts and kerbs, monumental and architectural granite and 'other').

The first two months of 1932 saw a massive rise in imports as foreign producers did their best to beat the tariff. But the big drop that followed was not sustained. Month by month, the value of imported monumental and architectural granite rose and continued to do so throughout 1933, until by July that year it topped the last pre-tariff month and climbed to a post-tariff record of over £14,000. As is so often the case with government measures, the tariff's actual effect was – well, not entirely as intended. What it did in the main was favour exports from Finland.

Major stone-exporting countries such as Germany found that the tariff made their exports less affordable in Britain, just as the Government wanted. But in Finland, variations in exchange rates outweighed this effect. The 20s of the British pound were worth 23s 6d in Helsinki. By enjoying the dual advantage of a cheap currency and low wages, Finnish exporters were able – in Diack's words – to 'snap their fingers at the tariff barrier'. As a result, Finland picked up the trade lost by their priced-out competitors, and since they were well able to cope with the extra volume of business, the tariff's net effect on overall British imports was almost nil. By July 1933 Finland was supplying forty per cent of all imported tombstones, with Germany second, Czechoslovakia third. And, in a turn-up for the books, the first ever consignment of tombstones – one solitary ton – arrived from Soviet Russia. Clearly, a fifteen per cent tariff was not enough to deter imports from all countries.

Also there was the embarrassing fact, not often articulated by the British quarrying industry in its meetings with Government, that because it was frequently unable to satisfy all the orders it received, it was forced to buy in from elsewhere. British quarry companies actually *needed* those hated imports. Britain was Finland's main export market, even for rough quarry blocks – a large proportion of which came ashore in Aberdeen itself. The exotic new colours and textures of the foreign stones were popular with architects and their clients, as well as monumental masons and theirs. Moreover, old-fashioned restrictive working and employment practices, reluctance to invest in newfangled machinery and a general lack of fresh recruits willing to take on a life of hard manual labour like their forefathers, were all gradually working against the old home advantage.

While complaints continued about the 'dumping' of tombstones from Finland on the British market, former export markets like the US dried up, and certain London boroughs began experimenting with pavement setts and kerbs from Mysore instead of buying British. Importation continued: from Finland, Sweden, Norway, Germany, Italy, Czechoslovakia, the US, Yugoslavia, Denmark, Belgium, South Africa, the Channel Islands, the Soviet Union and Austria. Yet, in all the figures published by the Board of Trade during the tariff fiasco, only once do any imports from British India appear. That was in January 1933, when a consignment of setts and pavement kerbs weighing 759 tons and worth £2,805 landed at London docks – destined no doubt for those traitorous metropolitan councils.

Diack wrote:

> . . . they are being offered at a price with which the British owners are finding it exceedingly difficult to compete. Indian kerb is being quoted at 2s 9d per foot run, but . . . owing to the way it is dressed it is worth about 9d per foot more than either Norwegian or British. In other words, Mysore kerb is of good quality granite and

well dressed. But the Indian kerb-dressers are paid the equivalent of 2s or 2s 6d per day. It is not surprising . . . that owners are viewing with some unease the inrush of kerbs produced by coolie labour under conditions that would not be tolerated for a day in this country.

Apart from the language Diack used to express it, one might say *plus ça change*.

There is no doubt that stone imports from India had begun at least a decade before William Bowen died, as adverts placed in the 1911 *Stones Trades Journal* by the merchant Gulamali Gulamhusen of Bombay (est. 1888), show. As early as 1917, reports from Aberdeen complain that: 'The monumental trade is largely dependent on foreign granite.' By 1921, the year before William's death, William Diack was writing: 'Granite headstones of foreign manufacture have recently been offered in this country at prices with which it is impossible for local manufactures to compete.' Yet a debate at the annual meeting of the National Association of Master Monumental Masons, reported in October 1923, was vociferous about imports from Scandinavia and Germany, yet made no mention of India at all.

I visited Mossfords' workshops to see the elements of William's tomb being bathed in acid and repolished, and his inscription re-enamelled. While I was there, Simon Morgan allowed me to read through Mossfords' company minute book covering the period April 1922 to August 1934. These made frequent mention of the chairman's absence on business in Spain and Italy, where he was negotiating with Italian quarry owners working Carrara Marble, and Spanish ones at the Ferrol Granite Company Ltd (in which Mossfords owned shares, and to which it made loans when business was poor). However, never in twelve years did the neat hand of the company secretary record the word 'India'.

All of which leads me to conclude that Anglo-Indian trade in mon-umental granite, worked to designs supplied from the UK, was not established as early as the 1920s. Even as late as the 1930s, Indian granites were satisfying UK demand only at the bottom end of the market. So perhaps my great-grandfather's granite did, after all, hail from Aberdeen. Or possibly Finland.

The situation today is transformed. Europe remains a major stone producer, 80 per cent coming from countries in Southern Europe: marble from Italy, Spain and Greece, slate from Spain and Portugal. In 2005 the European Union recognized 60,000 firms working orna-mental and dimension stone, employing about 600,000 people. But for granite, the contrast is stark. In 2010, the latest year for which figures are available, UK quarries raised about 367,000 tonnes of granite, nearly all – almost 271,000 tonnes – crushed for roadstone. We exported only 1,450 tonnes of worked stone in total. By contrast, in 2010 UK imports of all building and dimension stone totalled just over one million tonnes.

Three years earlier, China was exporting nearly 6.5 million tonnes, way ahead of India's nearly 2.5 million tonnes. Indonesia and Brazil came third and fourth that year; Italy, home of Carrara Marble, fifth at under a million tonnes exported. In these world compilations, the UK does not figure as a stone exporter. British monumental masons tell the same story. If one were to commission two similar tombstones, one in Chinese granite and another from any British producer, of any rock you care to mention, the Chinese one will arrive more than twice as fast, at less than half the price.

It is hard to argue against prices per tonne. Taking all rock-types together, stone imported from Italy in 2011 cost on average £990 a tonne, whereas stone from China cost only £298 – and from India £223. Driven by price, importers shift to cheaper sources; and yet, despite falling prices, in 2008 the UK spent a new record sum of

£300 million on imported architectural and ornamental stone weighing over 100,000 tonnes. According to the British Geological Survey's latest figures (2010), tonnage exports of worked and unworked granite from the UK have fallen to a mere 4.81 per cent of our imports. In tonnage we import twenty times more granite than we export, most of it (63 per cent) as worked stone – for shopping centres, kitchens, bathrooms and, when it's all over, our headstones.

We are now too sophisticated to dig our own rocks, but nevertheless we still face our buildings, furnish our kitchens and cover our dead with fancy stone – and we even save money doing so, thanks to cheap foreign labour and low-cost petroleum that has brought us the blessing of underpriced transport. In return, we export our new cargoes, the products of our indolence and complicity: greenhouse gases, poverty, water shortage, child labour, silicosis, unsafe working conditions and poor environmental practice.

Investigations in India by environmental groups and the BBC have revealed workers, without protection, splitting rocks with hammers and chisels and being paid around £1.50 a day for their trouble. Many have turned out to be farmers driven from their farms by drought, which water-hungry quarrying and stoneworking only make worse. Old women and children often work as bonded labour, in illegal quarries opened without planning permission. In the desert state of Rajasthan, in north-west India, the town of Kishangarh has a water-supply problem exacerbated by quarrying and stone-finishing, which are its main industries. The nearby quarries, which include the source of the famous Makrana Marble used to build the Taj Mahal, send blocks to Kishangarh, which boasts many hundreds of cutting, finishing, polishing and sculpting firms employing about 100,000 people. Their workshops draw stone from quarries and mines within a radius of up to 300 kilometres, and saws and polishers in just one such factory can easily consume 10,000 litres of water a day – all of which is now abstracted from groundwater at a rate faster than it can be replenished. Fine dust from these factories cakes the landscape around and, local farmers say, prevents what little rainfall the area receives from penetrating the ground. While in Aberdeen water has to be pumped away, in Kishangarh two large lakes, which local environmentalists say would once keep the town watered even when the rains failed, have vanished in the last two decades.

India's booming stone trade is being driven by rising domestic demand, as the country's burgeoning middle classes spend their money developing their dwellings. But India's export trade is booming too. Towering container vessels furrow the southern oceans, staining the skies with diesel smut to bring us these treasures. These distant oceans and their wide skies may be out of our sight, but not to satellites.

Since 2001, satellites have been able to observe trackways across the oceans, lines of towering cloud produced by ships' exhausts, their carbon-particle aerosols rising up, acting as nuclei for cloud condensation. Newer satellites, such as the Dutch and Finnish-built Ozone Monitoring Instrument (OMI) aboard NASA's *Aura* satellite, can see nitrogen dioxide in the atmosphere. They show these busy shipping lanes even more clearly, especially across the Indian Ocean, joining Singapore and China with Sri Lanka, from Aden to Suez, stretching across the Mediterranean from Alexandria to the Strait of Gibraltar. Rounding Biscay and Finisterre and gingerly threading the eye of the English Channel, at journey's end these ships – whose funnels have streaked the atmosphere around half the globe – inch into bleak havens at Rotterdam, Felixstowe and Tilbury with their cargoes of cobble setts, kerbstones, worktops, gravestones and cladding slabs.

Since I started writing this book, I have gained the impression that every geologist old enough to have bought a house (or a tombstone) can tell a story about this. The setts recently used to pave the

Annenberg Courtyard at Burlington House, London, where I work, derive from several quarries in the UK, mainland Europe and China. However, the quadrangle of Cardiff University, my cost-conscious alma mater, has been even more recently relaid (according to a member of the current professoriate) in granite setts, all of them from China. And one distinguished academic from Princeton, US, has told me how she relined her bathroom in an attractive limestone that turned out, surprisingly, to be British. However, from its origins in some UK quarry it had been exported en bloc to China, where it was sawn and polished before being re-exported to New Jersey.

Mossfords completed the restoration of William Bowen's tomb just a few months before my father died, which meant that the niche for cremated remains that I had had installed in its new concrete base-plate would be needed rather sooner than I had expected. It also meant that I was faced with the task of erecting a headstone to carry my parents' names and the names of my two second cousins, whose deaths in the Depression years had gone unmarked all this time.

Needless to say, the stone had to be three things. The inscription, to match the others, had to be in Welsh; it had to be made of grey granite, and that granite had to come from somewhere in the UK. Simon Morgan is, at the time of writing, following up three possible sources, two in Scotland and one in North Wales. He has prepared me for a long wait: sourcing ornamental stone in the UK is no easy matter. In the spirit of research, I paid a visit to a swanky London stone merchant not far from where I live, and pretended I was interested in a new kitchen worktop – particularly if it could be made from locally sourced rock. Clearly, the question had never been put before, so I explained that I was a geologist and therefore peculiar. The assistant showed me samples from France, Spain and Italy, but didn't think he had anything from the UK.

I asked him if he had heard of any move to introduce some form of 'rock miles' labelling on his products, in imitation of the 'food miles' scheme. He laughed and showed me a pink granite from China. I cannot now remember the precise trade name under which he was selling it, but it strongly recalled those other marketing coinages, such as 'Balmoral Red', given to imported rock from Finland back in the 1930s. This granite was, he said, one of their best lines, combining attractive colour and texture with a competitive price: just the thing for chopping your farmer's market produce on with a clear conscience.

I went back to Rubislaw Quarry – or 'Rubislaw Loch', as it is apparently becoming known – in the company of its new owners, Sandy Whyte, a semi-retired oil industry consultant, and Hugh Black, who is retired from the construction industry. I asked them how two

Sandy Whyte (left) and Hugh Black, new owners of Rubislaw Quarry, Aberdeen.

Aberdonian lads in their fifties had come to buy Europe's deepest
man-made hole, the longest-lived quarry in the Granite City, with a
bottom below sea level and which now, after forty years of neglect,
had become drowned beneath 400 feet of water. 'It was Sandy's fault,'
said Hugh. They both laughed. Some wealthy men in their fifties buy
Maseratis, or moated castles in Burgundy. Hugh Black and Sandy
Whyte bought a hole, and did so basically because, at only sixty thousand quid, it seemed like a snip.

Aberdeen's granite buildings glistered in the sunshine between
showers on the day of my visit, which took place almost exactly thirty
years since I had first parted the undergrowth and goggled down into
Rubislaw Quarry. Granite's near indestructibility gives Aberdeen its
strangely paradoxical look of being at once ancient and yet apparently freshly minted, as though built yesterday. The gleaming
nineteenth-century facade of Marischal College (built to house
Aberdeen University, now leased to the City Council) had recently
been cleaned, enhancing this impression. As we drove past, Hugh
pointed out that, while the building's interior is of Rubislaw granite,
its famous pinnacled facade is of another famous Aberdeen grey, the
Kemnay granite – most recently used to face the beautiful, and mas-
sively over budget, Scottish Parliament building at Holyrood. Only a
politically sensitive site like that could have commanded the exclusive
use of freshly won Scottish rocks on such a scale.

Although quarrying might be all but dead now, resource extraction
still runs deep in the Granite City's psyche – rock from the Earth, fish
from the ocean and now, offshore oil and gas, which seem to combine
some aspects of both former staple industries. At first, neither Sandy
nor Hugh had any plan as to what to do with their purchase. They
bought it out of pure sentiment. Their desperate boyhood agonies
were played out in and around this very hole with its dizzying, sheer
walls. Hugh recalled, with a shudder, how one of his contemporaries

once clambered out along one of the steel cables, slung across from side to side, and dangled there hundreds of feet over the yawning chasm. From these steel ropes (a crane system invented at Kemnay, and named 'Blondins' after the French tightrope walker Charles Blondin, 1824–97), men were once lowered in and granite raised out, up to twenty tons at a time. During blasting, workers would be lifted up in the bucket – and since to remove them entirely would have taken too long, they dangled, at an assumed safe height, as the charges were detonated, protected by the steel bucket beneath them from any upwardly mobile projectiles.

The land surrounding the quarry had already been sold off years before Hugh and Sandy bought the hole itself. 'The risky bit, the bit with all the water in it, was the last to go,' Sandy told me. Other buyers had expressed interest, but only Hugh and Sandy's bid was unencumbered by caveats over planning permission or surveys. 'We said, "We'll take it, whatever",' explained Black. 'Because for a couple of Aberdeen guys like us to own such a huge part of the city's history was just . . . irresistible.'

Black and Whyte bought their sixty-grand prize. Considering that this huge hole is big enough to swallow Edinburgh's Castle Rock, and that it cost them only six times more, give or take, than it cost me to restore William Bowen's tomb, this conveyed to me an amazing sense of value. Oil company offices (including Chevron, where Sandy Whyte once worked, and ConocoPhillips) and a few modern housing developments now peep over Rubislaw's wooded rim. None was there in the 1980s, so since then the place has lost some of its eerie loneliness. But, just as when I first saw it, apart from those who live and work in the new buildings, nobody else would know it was there.

One of the first tasks facing the new owners of this fragment of forgotten landscape was to repair the boundary and install a gate somewhere. Rubislaw has just one accessible piece of shoreline, and

I followed the pair to it, past the dissuasive notices and barbed wire that reinforce the formerly ramshackle fence, over the old trees' talon roots and down the other side via scrambling nets pegged to the ground.

A plastic pontoon jetty jutted into the still, black water, to which was tethered a fibreglass elfin pinnace (dubbed the *Deepwater Explorer* by Hugh and Sandy) and a buoy from which hung, they told me, a submersible pump. Worried that the water level still seemed to be rising, the lads had decided to try to pump some of its six or seven million cubic metres away. This was partly inspired by caution over possible legal claims; but a plan was beginning to form in their minds.

'The quarry's been ignored for forty years, and we thought maybe there is a chance to do something here, because if we don't, there's another generation gone,' said Black. He and Whyte have since been vigorously interesting local children in their city's industrial heritage and are convinced, by their overwhelmingly enthusiastic response, that there is potential for some kind of heritage centre, to explain the Aberdeen granite industry to future generations as well as provide a focus for tourism.

There was a grey band like a tidemark all around the lake, much as you might expect to see around a reservoir in summer. It was perhaps four metres broad. The branches of trees and bushes in this zone were festooned with tattered skeins of dead grey algae. This lowering of the water level had been the result of just a few weeks' pumping at a modest fifteen litres per second – not enough to make any perceptible difference to the flow rate in the Rubislaw Burn into which they drained it. In fact, barely a weekend after the pump was first turned on, Hugh and Sandy went back and could hardly believe how much they had managed to achieve in so short a time. To their satisfaction, this suggested that water recharge to the hole was manageably slow.

Sandy and Hugh now want to interest Aberdeen's universities, council and local industry in an ambitious two-stage plan to open the site to the public. Beginning with an education and heritage centre, combining industrial archaeology, ecology and geology with commercial activity (a business and conference centre) to make it commercially viable, they would then develop the site further as an outdoor activity centre. Chartered architects and planning consultants Halliday Fraser Munro have devised concept drawings for a striking signature building, as jagged as the cleaved granite itself, cantilevered out over the quarry – the drop enhanced by further lowering the water to a broad ledge a hundred or so feet below. This ledge would form the base for diving and climbing or any other activities. 'We could even use the loch as a refugium for the Arctic char,' says Sandy. *Salvelinus alpinus*, a highly flavoursome relative of the salmon, is one of the rarest fish species in Britain. It is found naturally in deep, cold lakes, mostly in Scotland, but it is currently at risk from acidification.

Of all the schools he has visited, Sandy recalls one in particular – Kincorth, set amid the last city council-housing estate to be built from local stone. 'We could tell the children that they could look out of any window and see Rubislaw granite. Literally, it brought it home.' No memorials last for ever, except the generations who remember.

This is the past's lesson for the future. Resource exploitation is, always, a necessary but necessarily temporary use of land. The oil business, which not only rescued the UK economy but, specifically, Aberdeen just as quarrying was dying in the 1970s, now peers down into the depths of Rubislaw's black waters and sees what will one day become of it.

Aberdonians have found, in Black and Whyte, the intermediaries they need to help them commune with those heroic forebears who cleft the granite, rock of ages. As for mine, William Bowen's restored

tomb, mixing the granite with the Pennant, shines brightly from its
select location on the slopes of Mynydd Merthyr, across the parking lot with which they paved his portal to paradise, over Afon Taff,
to a desolate patch of green where his vanished colliery once drew
forth past richness; richness with which he and all his comrades
bought their dignity in life and continuance in the memory of grateful heirs.

The restored tomb looks down on the site of the former
Merthyr Vale Colliery, where William Bowen's working life was spent.

5

Oracle

And what news from the kingdom of subterranean darkness and
airy hope? What says the swart spirit of the mine?

Sir Walter Scott, *The Antiquary*

Homer's *Iliad* and *Odyssey* fossilize the thought processes of Bronze
Age people, including their belief that the past lives all around us and
that ancestral shades can be invoked, communicated with and asked
about things to come. This sense of the past living around us clearly
seemed strongest to those ancient peoples in places where the underland is physically accessible – at portals like caves, grottoes and even
man-made pits. These gateways to the netherworld became sacred
places where the very stones could speak – '*te saxa loquuntur*'.

In our time, the scientific discourse with rocks that has been going
on for about three centuries has brought us economic benefits and
comforts that we all enjoy, and without which we could not survive.
Those centuries have also fixed in our modern minds a fundamental
cultural construct with, at its root, two often conflicting themes.
The key ideas, first formulated in the Romantic movement, are the
worship of nature and the glorification of the individual. Although
Romanticism and industrialization have often fought one another,

in fact they sprang together from the Enlightenment like conjoined twins.

A long-standing association between the underworld and certain privileged mortals – the priests and poets who share an understanding given otherwise only to gods and the dead – has not gone unremarked by generations of geologists. In building their own mythology (as they tend to do in the introductions of undergraduate textbooks, for example) geologists today cast themselves in the role of latter-day prophets; savants who have at last learnt to decipher the hand in which the story of the past, and its predictions for the future, are written – the true and original language in which 'rocks speak'. Geologists therefore tend to recruit into their pantheon many of the mortal ancients, just as all new cultures, with new systems of worship, appropriate the deities of old.

One of those honoured in this way is the Greek geographer and historian Strabo (c. 63 BC–AD 24), author of seventeen volumes known as the *Geographica*. He is revered for making many pioneering observations about the Earth and its deep history, especially his attempts to explain how sea shells come to be found in rocks now lying far above present sea level. As the great nineteenth-century geologist Sir Charles Lyell noted in his *Principles of Geology* (1830–32), Charles Darwin's bunkside reading aboard HMS *Beagle*, Strabo was particularly prescient about the tendency of certain portions of the Earth's crust to rise, and others to fall, under the influence of geological – and especially volcanic – forces.

Strabo travelled widely in compiling his highly descriptive and historical book, which presents a detailed and surprisingly accurate picture of the antique world and of how that world 'appeared to itself'. When describing the city of Smyrna, for example, this prototype Baedeker does not fail to mention the *Homereum*, a quadrangular portico containing a shrine and wooden statue of

Homer, for, as he writes: 'the Smyrnaeans . . . lay especial claim to the poet'. This tradition, of which my father had been aware back in 1968, held that Homer actually composed his epic poems in a nearby grotto, which has ever since been held sacred to his memory.

Legend, here, is perhaps more important than literal truth; it tells us that, to the ancients, even a blind poet like Homer was expected to retreat to the gates of the underworld, to receive what Cicero described in his *De Natura Deorum* ('On the Nature of Gods') as 'divine afflatus' – the unexpected inrush of breath; literally, inspiration – that could leave the poet senseless or in a trance-like state, stunned by this touch of the divine. Richard Gregory, then editor of the journal *Nature*, conflated science and the divine when he wrote, in 1928: 'The conviction that devotion to the study of nature exalts the Creator gives courage and power to those who possess it; it is the divine afflatus which inspires and enables the highest work in science.' Whether they connect their feelings to the supernatural or not, few Earth scientists would deny that those moments of vision, achieved in solitude while contemplating the rock faces of a quarry, number among their most transcendent.

All around the Mediterranean Sea there were, and still are, many portals to the underworld, because the countries surrounding it are built largely on limestones that were laid down in an ancient ocean embayment known as the Tethys. Tethys was an arm of the global ocean 'Panthalassa', separating the northern and southern lobes of the supercontinent 'Pangaea', and which, long before the Atlantic opened, stretched from what is now China to what is now America. The sediments deposited on its floor were folded, faulted and uplifted by the collision of Africa (which formed part of Pangaea's southern lobe, Gondwanaland) with the northern lobe, Laurasia, which included Europe and Asia.

In Europe, this collision eliminated old Tethys during the formation

of the Alps, and the same continuing collision is today gradually elim-
inating Tethys' modern successor, the Mediterranean. The geological
underpinnings of Mediterranean countries ('cradles of civilization' as
we used to call them when it did not seem incorrect for Europeans to
value their own culture above others) consist therefore of a potent
combination: soluble limestones – with their natural tendency to form
extensive cave networks – and active tectonics.

Most faults that geologists map as they trace out the pattern of the
rocks on the ground are dead, extinct, their active life long past. But
some live still, and move during earthquakes that happen today.
Because the whole Mediterranean region is tectonically active, it
boasts a large number of active faults, some of which even break the
land surface and leave visible scars. Nor is it unusual for such faults
to occur close to mineral and thermal springs, exhalations of subterranean vapours and even volcanic vents. Small wonder, then, that
places like this (so alien to British eyes, accustomed to the stiller landscapes of our relatively quiescent underland), places where alarming
and unusual phenomena have manifested themselves strongly or in
combination for millennia, have been imbued with religious dread
since antiquity and associated with supernatural beings.

These localities, with their deities – especially Gaia (the Earth
Mother) and Apollo (son of Zeus and Leto and god of prophecy,
among other things) – often became places of pilgrimage. Many, like
Delphi in Greece, developed into rich institutions where communion
with the gods was mediated by a cadre of professional priests and
oracles. During the nine days of the year when the Delphic oracle was
open for business and taking questions from supplicants with the
means to support their consultation (which were made by city states,
monarchs and imperial tyrants as well as rich individuals), these
intermediaries would retreat underground and then emerge, after
their period of holy communion, with the requested prophecy.

The age-old association between active geological sites and ancient holy places can be useful to us today. Italian geologist Dr Luigi Piccardi, who works at the magnificently named Centro di Studio di Geologia dell'Appennino e delle Catene Perimediterranee in Florence, has been able to use sacred sites from antiquity to locate hitherto unsuspected active faults. Although these may have been quiescent for hundreds or thousands of years, and so long passed into the realm of the forgotten, they may still pose a hazard to people alive today, and it is important to know where they lie.

The connection between caves and gods was clearly a fixture of the classical and pre-classical mind. So, when in Homer's *Odyssey* the witch Circe sends Odysseus to the land of the Cimmerians to raise, and commune with, the shades of Tiresias and others, rich libations and the sacrifice of a ram were not considered enough on their own. Even in that gloomy land of the dead, communing with spirits and divining the future required the hero to dig a hole in the ground.

Greek oracles of the dead – *nekromanteia*, as they were first known – nearly always occupied entrances to the underworld, either natural or artificial. The isolation and darkness of a natural cave might even be enhanced by the construction of thick-walled, windowless chambers, in which general disorientation induced by darkness might be heightened by other means. The entrance to a sacred and typically underground chamber was called the *adyton*, root of the word 'adit', or horizontal access passage to a mine.

In the early 1970s, at a site near Ephyra in Thesprotia (western Greece, on the borders of modern-day Albania), archaeologist Sotirios Dakaris excavated a large complex of chambers surmounting a crypt cut into bedrock, where he found the remains of certain unusual foodstuffs, including significant quantities of lupin seeds and broad beans. Dakaris noted that both these, ingested, may induce hallucinations,

especially when green. In conditions of darkness and sensory deprivation, such chemical assistance may have been used to help bring on
the transcendental.

Not all underground sacred spaces needed such help. The name of
the most famous oracular site of the ancient world – Delphi, *ompha-
los*, or navel of the world – comes from the same word stem as the
ancient Greek word for 'womb'. Delphi was an Apollonian temple in
the classical period, but at its most archaic stage of occupation
(dating to the end of the last Ice Age) it was probably dedicated to the
mother of all Greek deities, the wide-bosomed Earth goddess Gaia,
guarded by a dragon serpent called Python. Apollo is supposed to
have slain Python in claiming this dramatic cleft in the slopes of
Mount Parnassus for his own.

Delphic divination was not assisted by hallucinogenic seeds, but by
mystic vapour – the *pneuma* – which was inhaled by the female oracle,
the Pythia, perched on her tripod over a fissure in the rock. Researches
now suggest that these vapours were themselves no myth –they were
earthly exhalations rich in the hydrocarbon gas ethene (or 'ethylene',
chemical formula C_2H_4, an intoxicative inhalant, and the same
substance responsible in tiny quantities for ripening bananas and
other fruit). This gas is released by seismic action along the active geological fault that runs beneath Delphi, through hydrocarbon-rich
sedimentary rocks deep below. A cave would have concentrated these
gases to near-toxic levels, quite enough to induce hallucinations and
trance-like states. While under the influence of the *pneuma*, the Pythia
would make her famously Delphic utterances, in the divine words of
Apollo himself.

Not to be understood by mere mortals, these were then 'interpreted' – translated into mortal tongue – by attendant priests, who were
supposed to have learnt the language of heaven and were able to tran-
scribe it into neat iambic hexameters. (On occasion, these hexameters

contained metrical errors. Since these lines were supposed to be emanating from the realm of perfection, this fact puzzled many even
among the ancients.)

Most other oracular caves in the ancient world relied simply upon
isolation and sensory deprivation, both of which can, on their own,
induce hallucinatory experiences. The Pythia may have been unusual
in having access to naturally occurring psychotropic gases, and perhaps the fact that Delphi could cook with its own gas lay behind its
prolonged success as a centre of divination.

But Delphi was not unique. As the scholar Yulia Ustinova has written in her book *Caves and the Ancient Greek Mind*, the precinct in
Hierapolis (in modern Turkey, near to natural hot springs) resembled
the Delphic sanctuary in two crucial respects: 'the connection between
the Temple of Apollo and a hole emitting gases, and the position of
the temple above an active fault'.

In seeking to explain the association between holes in the ground
and oracular sites, Ustinova invokes the Greek notion of ultimate
truth. The Greeks believed that such knowledge, being the preserve of
the gods, could not be understood by mortal men (perhaps an early
version of British biologist J. B. S. Haldane's famous speculation that
the universe might not only be queerer than we suppose, but 'queerer
than we *can* suppose'). In order to partake of such knowledge and
receive the divine afflatus, mortals would have to liberate their souls
from the burden of the mortal body.

This they did by attaining a state called by various Greek names
that all provide roots for familiar words in English today – like *eksta-
sis*, *mania* and *enthousiasmos*. Such states descended as the mind
became possessed by a super-being, and the result could be prophecy,
poetry and insight. This idea, of holes in the ground peopled by
priestly intermediaries full of enthusiasm and even ecstasy, offering
interpretations of the past that might also serve as predictions of the

future, to my mind presents a fairly apt description of a party of geologists in a quarry.

Through time, the association in the Greek mind between prophets, caves and Earth movements became more or less universal. The grotto was the place in the landscape most appropriate for divination; the blindness of the location either substituting for actual blindness, or (as with Homer at Smyrna) adding to the poet's own. During the subsequent centuries-long eclipse of the classical world, the same holy dread of sacred places, close to the portals of the underworld, where the dead and the immortals dwelt, survived – however dimly – into medieval Europe. Here it waited to be rediscovered in the eighteenth century, when a return to classical forms and ideals heralded the arrival of the Enlightenment.

The nature of our engagement with holes in the ground inevitably changed as economies became more industrial and societies more complex. Most of us can witness this process of societal evolution simply by tracing our own family timelines and discovering the occupations of our ancestors. At some point in history, everyone's ancestors were farmers. On my mother's side, the last generation to work the land was my great-great-grandfather George Bowen, William Bowen's father. Born in East Buckland, Devon in 1830, he was listed in the census as an 'agricultural labourer'. Farming last appears a generation further back on my father's side; they lived at Risk Farm, near Mossley, and moved to East Manchester, where industrialization became advanced at least one generation earlier than it did in South Wales.

Subsequent generations on both sides, however, were engaged in a mixture of mining, the fashioning of stone as monumental or architectural masons and the smelting and casting of the metals derived from it – the winning of Earth materials and their transformation. That is what underpins all advanced societies, and always will. Only when I come to my maternal grandfather (and, on my father's side, to my great-grandfather) do I find the family breadwinner earning his crust indoors, in a shop, office or classroom. It is important, though, not to be fooled by this (and I believe that many are) into the logical fallacy that the diversification of our society and economy away from basic activities means that farming and mining are dispensable. Someone, somewhere, must do these things.

As the human need for metals and sources of concentrated energy multiplied at the end of the eighteenth century, man-made holes in the ground proliferated in the landscape. Holes now became places not so much of religious or mystical power, but of daily work – worldly, rather than otherworldly. They still mediated the exchange of knowledge and power, but of a different kind. The knowledge required was practical – to enable us to sink mines successfully and operate them profitably.

It is hardly surprising that the nineteenth century became the golden age of geology. It was the first century in which people visited the underworld in force. For all that geology's roots lie in the industrial mining that began in earnest during the eighteenth century, until industry gathered pace most people continued to eke out their existence, scratching the surface of the planet and contenting themselves with the level of energy input that could be derived from sunlight acting on this year's leaves, or the short-term energy reservoirs built up in wood and peat. Tapping into energy concentrated in deep time meant ever more digging and delving, and ever more planting of large infrastructure far down into the landscape.

The first literary works dedicated to the art of mining were written in the dying years of the Middle Ages – the first being Georgius Agricola's *De Re Metallica* ('Of Metals') of 1556. Miners, with their special knowledge and closed – or at least close-knit – communities, never lost their mystic aura. Today, we ascribe a particular meaning to the word 'occult', implying magic and superstition. But 'occult' only means 'hidden', and in the medieval world, where all knowledge was closely guarded by those who possessed it, no distinction existed between what we would call 'magic' and 'science'. All special knowledge was secret. The distinction between *truly* secret knowledge (which is authoritarian and jealously guards its mystique largely because it is otherwise ineffectual) and knowledge that is open to all (because it is freely demonstrated by experiment) – did not exist. But all knowledge means power, especially knowledge that *works* (which is a remarkably precise way of defining what we now, in English, mean by 'science'). It is hardly surprising, then, that those who possessed it at first mimicked the magicians in preferring to keep it to themselves.

The latter years of the eighteenth century were a pivotal moment at which a Western culture founded on centuries of landowning and farming, impregnated by the rediscovered classical world and its spirit of open inquiry, found itself pregnant with both industrialization and Romanticism. Demand for useful knowledge – needed to support the growing new industries – outstripped supply and generated a new market for men of genius or 'adepts' as they were called. Working at the limits of the known and often mixing science, folklore and superstition, such men emerged to advise those who, as owners of land or money or both, wanted to know how they might best speculate to accumulate.

This was the age of Figaro, as depicted in Beaumarchais' play and Mozart's opera; the low-born man of talent who chafes in his servitude and yearns to be free. Industry opened up new ways of making money. Landowners wanted to know how to get more value from their acres than by grazing sheep or growing turnips. The first industrial pioneers, cash-rich entrepreneurs, were eager to sink their capital into new commodities for the emerging markets. But with what, and how?

As I found in my brief career as a geological consultant to oil companies, consultancy is a curious occupation, and involves persuading clients to part with cash for the benefit of your expert opinion and advice. To be successful in this, an air of confidence is essential, but a fine balance has to be struck. Should you present yourself with too little confidence, you will fail to inspire it. Show too much, however, and you risk coming over as a quack.

Nowadays, professional consultants can call upon an armoury of confidence-boosting paraphernalia – university degrees, professional accreditation, and so on. But the late eighteenth century had none of this – there was only charisma, reputation, track record, word of mouth and cheek. As the quality of advice was inevitably much lower in those days, a reputation of the wrong sort was easy to acquire, and dogs with bad names tended to keep them. It was important to be mobile, so as to keep ahead of any bad news. Pioneers of the geological profession may have been savants of a sort, but they were not men of leisure. Unlike their more patrician 'learned' counterparts, they needed the money. And because other men's money was riding on what they said, stakes were high.

Entrepreneurs and landowners needed these men – and almost certainly, for that very reason, reviled them, never certain whether they were just throwing their money away on someone whose advice might be no better a guide than their own instinct or a well-aimed prayer. I experienced this first-hand once when I was sourcing a suitable stone for an engagement ring. As I was already married to the intended recipient, the purchase was a little belated; but I had been unable to afford one at the proper time and had issued a verbal IOU. Finding myself in Sri Lanka, fabled source of gemstones throughout history, I decided to talk to some artisanal miners.

The stones that these men find derive originally from very ancient rocks, exposed in the heart of the island. They have either been

weathered out from the parent rock and sit around in the thick tropical soil, or they have been washed out from there and transported,
down a river laden with yellow sediment, and dumped – maybe also
concentrated with others – in river floodplains. Sometimes, a pit dug
into these deposits will hit a thin seam of gem-rich gravel, sometimes
not. Much, much more often not. Sadly for the prospectors, there is
no reliable 'scientific' method of predicting exactly where, in these
piles of complex river-laid sediment, pay dirt is likely to lie. So I was
interested to find out how they chose their locations.

They were not secretive about this – quite the reverse. All of them
eagerly explained through my guide that before reaching for their
spades, they employed complex, picturesque (and not inexpensive)
religious ceremonials to help them pinpoint the most propitious sites.
They were evidently proud of this, and the larger gemstone shops on
the island usually have visitor centres where you can see photographs
of the great mystery of divination taking place.

This method is every bit as good as any other; in fact, it is better,
since it makes a pretty tale with which to enliven and prolong the visitor experience. Aside from that, however, it is also true that they
would do as well to save the priest's consultancy fee, close their eyes
and throw a stick. But it is not human nature to admit, quite so
explicitly, that mere chance rules one's life, even on an island whose
ancient Persian name of *Serendib* ('the place where one finds that
which one is not seeking') gives us the word 'serendipity', or happy
accident. In the end, I procured an aquamarine, which they cut and
which I later had set into a band of ancestral gold, the melted-down
wedding rings of William Bowen and Margaret Rosser.

People have little choice but to forgive the gods when they come up
short, but such generosity of spirit rarely extends to the secular
consultant. One senses this in the way that eighteenth-century industrialist Matthew Boulton (1728–1809) sneeringly referred to the

engineering geologists advising him on canal building as a 'tribe of jobbing ditchers'. Holes in the ground were acquiring a new kind of intermediary; one who, because he claimed a rational basis for his advice rather than the capricious inspiration of the *pneuma*, could be blamed if it all went wrong.

Geologists, especially those who work on the engineering side of the subject, quickly made a demigod of William Smith (1769–1839), who had to fight many battles while trying to earn a living from a fledgling science as an adviser on canals, land drainage, mines and quarries. While Smith's speculations were not always successful – one disastrous quarry venture landing him in debtors' prison – Smith's original scientific contribution has always lain beyond reproach or doubt. He may not have been immediately accepted by his London contemporaries, who formed their Geological Society in 1807, but he did draw the first geological map of any country based on strati-graphic principles, which he himself enunciated.

The first of these principles states that unless rocks are deformed, older ones will be found on the bottom and the younger on top. The second states that sediments of equivalent age, independent of rock-type (which is governed by conditions of deposition), may be identified as such, and correlated across country, by the fact that they will contain the uniquely distinctive fossils of their time. These ideas, combined with the visual language of the geological map, made Smith's name, and he attained thereby the highest honour for any scientist: his work became so much part of the furniture that people today can hardly imagine that such central tenets even required invention.

To give them credit, when a delegation from the Geological Society of London first saw Smith's map, they immediately recognized its worth. Indeed, they thought it too important to be left to an untu-tored 'mechanick' like Smith and bought a copy to use as the basis for a map of their own. Despite being the most advanced group of thinkers on Earth science in the world at that time, and despite their commitment to 'fieldwork first, speculation later', not one of them had had the idea of simply finding out what rocks underlay the landscape and of plotting this knowledge on a map. Advanced thinkers or no, none of them needed to earn his living by geology. Ask any practical geologist today where every question about mining, quarrying or civil engineering begins, and they will tell you – the geological map. The consequences of not consulting one of these distillations of what the rocks are saying are what Gillian Shepherd MP really meant when she suggested that, to an MP at least, geology equals trouble.

If Smith were what philosopher Isaiah Berlin would have termed a hedgehog (a thinker who sees the world through the lens of one or two big ideas), an early geologist in the generation before Smith was the fox who knew many things. Venturing even earlier into this risky new profession-cum-priesthood, always on the fringes of respectability, came Rudolf Erich Raspe (1736–94), with his pointed features, red hair and nervous, rushing gait. Although both were men of the eighteenth century, Smith, by the simple, protean grandeur of his achievement, was more Beethovenian than Mozartian. Smith's story was more *Fidelio*, whereas Raspe had much in common with *Figaro*.

Charles Lyell did his best to resurrect the name of Raspe in his *Principles of Geology*. While preparing the first edition, Lyell entered into correspondence with one James Hawkins who – by then himself a Fellow of the Royal Society – had befriended Raspe in his most desperate hour and become his pupil. In reply, Hawkins described his former mentor in fond and glowing terms, saying: 'He possessed a more extensive knowledge of any man I recollect.' But of those who know the name of Raspe today, few remember him for great knowledge, or even know that he was an early geologist. Most know him only as one of his century's greatest scoundrels.

When Hawkins first met Raspe in London in 1775, the older man, Raspe, was embroiled in a scandal that had driven him into exile. Shunned by his learned contemporaries and thrown upon his own resources, he used his extensive knowledge to make money in any way he could, a whole generation before Smith faced the same problem. This hardship brought out the best in him, though at the price of academic success and subsequent fame. The mercurial, opportunistic Raspe, despite playing a vital role in advancing our understanding of how the world works by internal heat, nevertheless sank into obscurity and became part of geology's forgotten underworld.

Raspe was born in 1736 in Hanover, Lower Saxony, to this day a staid city surrounded by peat bogs and pine forest under the grey North German sky. By virtue of England's Act of Settlement 1701, which decreed that no Catholic could accede to the British throne, the ruler of Lower Saxony – known as the 'Elector' – was to become the first Hanoverian monarch of England. This happened because his mother, Sophia of Hanover, was the most senior Protestant descendant of James the First, and so it fell to her son Georg Ludwig to assume the title George the First of England and usher in the Georgian age.

Raspe's father, Christian Theophilus, worked in the Hanoverian department of mines and forests as an accountant; an industrious and dutiful man well suited to the stuffy, patronage-driven, deferential and mediocre state he served. He hoped that his son would build upon his own career and sent him to study law at the then new University of Göttingen. But young Raspe's imagination had already been captured by conversations he had overheard between his father and the mine managers and engineers of the Harz Mountains, who visited him in Hanover to discuss their accounts. The Raspe family had taken their vacations among the miners, who had become his father's friends, staying at Clausthal, capital of the Harz, as guests of the so-called *Berghauptman* or 'mountain captain', Herr von Veltheim.

And so the young Raspe came to know from the inside the closed world of Europe's longest-established mining community, which from medieval times had fostered the impression of being the descendants of the dwarves of legend, custodians of mysterious treasures. Raspe descended with them into the underworld of the Harz, then the most technically advanced mining area in the world, seeing for himself the caverns they had wrought, how they used underground water to pump, ventilate and power their mining operations, and the way they divined the lay of the silver, lead and copper ores that had made them and the Elector of Saxony rich.

The Harz did more than throw a financial lifeline to the decaying monarchy it served. In an age that was coming to regard remoteness and wilderness as ideals rather than inconveniences, the Harz – with its colourful folk tales, Spectre of the Brocken, occult goings-on and hidden treasure deep underground – attracted the coming men of the Romantic age. Chief among them was the poet Johann Wolfgang von Goethe (1749–1832), a man would could reasonably be described as the co-chief architect, with Jean-Jacques Rousseau, of the entire Romantic movement. He was then working as a state official in charge of mines, and visited the Harz in 1780 to open a new one at the spa town of Ilmenau. (Geologists might think they recognize this name from the mineral 'Ilmenite', an iron-titanium oxide, though in fact this is a coincidence. The mineral actually takes its name from Lake Ilmen in Russia.)

At Ilmenau on 6 September, in a hut preserved to this day as a shrine to the event, Goethe was inspired to write the second *Wanderer's Nightsong*, famously set to music by both Schubert and Liszt, and thought by many to be the most perfect lyric poem ever written in German. A typical outpouring of the unquiet heart so dear to the Romantics, it tells of the solace to be drawn from communion with raw, untutored nature. Rugged mountains, hidden treasures, pine

forests, mists, medievalism and ancient legends of a hidden 'world
beneath the world' became the zeitgeist out of which scientific geology first sprang.

The young Raspe, eager for success in his father's world of favours,
patronage and preferment, became a highly catholic scholar, antiquary,
philologist, linguist and geologist. He rose quickly from clerk at the
University of Hanover library, to a post at the library of his alma
mater Göttingen, eventually becoming curator to the Landgrave of
Hesse-Kassel. In 1769 he was elected a Foreign Member of London's
prestigious Royal Society, and in 1771 he married Elizabeth Langens,
daughter of a wealthy Berlin doctor, who brought with her a generous
dowry.

On the surface, all seemed to be going well. But the first sign of
cracks appeared when the ambitious young functionary mysteriously
turned down a prestigious promotion, one towards which his career
had been apparently heading, namely to take charge of the Royal
Collection in Berlin. Raspe pleaded personal loyalty to his employer,
but this turned out to be every bit as fishy as it sounded. The real
reason he found himself stuck in Kassel was that he dared not
hand over the keys to the Landgrave's museum. This he had by
now laboriously catalogued, and in so doing, woven the rope that
would hang him. Should anyone decide to check the holdings against
the inventory, they would notice that a good many of the items so
meticulously listed had vanished.

Since 1769, the very year he was elected to the Royal Society, Raspe
had been steadily pawning his employer's possessions to help finance
the lifestyle that his elevated status demanded, but which his salary
was too meagre to maintain. Eventually the tangled scaffolding he
had constructed to support his deception and keep it secret, collapsed.
He was summoned for a formal stocktaking on 15 November 1774,
and later admitted: 'It might have been better had I not fled.' But flee

he did, leaving his wife, little son Friedrich, his position, everything –
except a few last trinkets of his employer's, with which he filled his
pockets to help finance the escape.

Uncannily attuned to the spirit of his age, Raspe had become the
very wanderer of whose unquiet heart Goethe had sung that night in
the forest above Ilmenau. Although he wasted much of his future life
fruitlessly grovelling for pardons that never came, or vainly seeking to
re-establish an academic reputation in a new homeland, he was now
completely released from the need for the patronage that had conspired with his weakness of character to ruin him. As his biographer
wrote in 1950: 'he at last discovered the courage he had always
lacked'.

Out of shame, disgrace and an understandable desire to remain out
of prison, Raspe fled to England, land of industrial opportunity. Here
he forged a new career, and so escaped the stifling, kowtowing atmosphere surrounding the effete princelings who ruled continental
Europe. While this new life brought little prestige, it did bring something precious beyond rubies. A thing of shreds and patches he might
be, but at least he was no longer beholden to a bunch of titled ninnies.
He could court the attention of the coming men; men with cash in
their pockets, the rising generation of industrialists whose ambition
would soon – literally – split the Earth asunder. He found self-respect,
working in many capacities – notably as a mineral assayer in
Cornwall, under the wing of the company run by Matthew Boulton
and James Watt.

Geology was only one of Raspe's many interests, but nevertheless
he deserves more credit than he tends to get for playing a significant
role in resolving a debate known as the 'Basalt Controversy'. This
matters, because the Basalt Controversy proved a crucial battleground
between two major schools of thought about the way the Earth
works, with far-reaching implications for this story. The debate

divided broadly into two camps: the 'Neptunists', who thought all rocks were produced by precipitation from primordial oceans, and 'Volcanists' who claimed that the Earth was a great heat engine whose internal fires drove geological processes, lifting and depressing portions of the crust.

Raspe gets credit here because he was responsible for promulgating the novel idea that basalts had once been molten, and for doing it so efficiently that he almost beat the scientist who first had the idea into print. Yet Raspe has been long cheated even of that achievement – which was undoubtedly because, at the zenith of his academic fame, he fell suddenly and catastrophically from grace. So complete was this fall that, as his biographer John Carswell wrote: 'Raspe's friends did their best to give the impression either that there had never been such a man, or that they had never known him. His memory meanwhile was kept alive by two other groups – his personal and literary enemies, and the police.'

The oddest thing, however, about this romancer of the stone is that despite a disgrace so complete that his scientific reputation did not even begin to recover until 1970, for over two centuries speakers of almost every major language on the planet have treasured one of his books. And since that book, and the immortal comic character it introduced, were dashed off just to pay his tailor's bills (and, like his other non-scientific literary works, appeared anonymously) we can say truly that scandal was the making of him. Had Raspe not been cast out from the academic paradise he so craved, the world would never have made the acquaintance of Baron Munchausen.

One sees in Raspe the combined force of the Industrial and the Romantic revolutions. One might think that the Gradgrind industrialist, in his ghastly parvenu pile built far from sight of the industrial squalor that paid for it (though this cliché has many exceptions), finds little in common with the poet starving in his garret. But both can qualify either as 'hero' or 'genius', two Romantic notions that both date from the late eighteenth century. Both industrialist and poet were fighting a common enemy – the same enemy targeted by Figaro himself – namely, the old, regulated society of kings and priests. What they struggled for was the right to self-determination. Against that backdrop, industrialization and romanticism complement one another perfectly, even as the new industrial labouring class flocked to buy the works of Lord Byron and Sir Walter Scott that poured off the new steam-powered presses. Industrialization and romanticism may have indulged in mock battle, but in reality both had sucked at the same teat – the past.

When Raspe's tale came at last to be properly told, nearly two centuries after he vanished into oblivion, it was largely by academics who, while rehabilitating him, nevertheless tacitly subscribed to the notion – unique to their tribe – that to be denied academic success is the ultimate failure. Even his would-be rescuers (unlike his biographer, whose work preceded theirs by over twenty years) took the view that turning knowledge into cash bore the stain of trade like the mark of Cain, and was a source of further disgrace to their hapless subject. I think this is a mistake, and to understand why, we need to consider the differences between how the spirit of the late eighteenth century manifested itself in England, as opposed to mainland Europe.

Britain, under the domination of England, has never been a sophisticated culture. We do not really do finesse. While we envy it and occasionally attempt to imitate it, French court shoes tend to pinch our big, flat British feet. 'Our Island Story', as it used to be called, has always been one of buccaneering brigands and pirates; the last barbarian invaders from the north lands, violent, and frequently – possibly permanently – inebriated. The roots of this unruly, rebellious character run deep, as the Cambridge historian Alan Macfarlane pointed out in his classic 1978 study, *The Origins of English Individualism*.

English social institutions evolved differently from those on the Continent, a process that can be traced back to at least the beginning of the thirteenth century. The British monarchy, unlike those elsewhere in Europe, failed to centralize and reinforce its power and so never became absolute. The English monarch was held in check – most famously by Magna Carta (1215), the 'Great Charter of the Liberties of England', the most significant early milestone in the centuries-long process that led to what we sometimes refer to as the 'rule of law'.

English property and labour rights remained vested in individuals, who could work on any land, trade freely and bequeath belongings to anyone they chose. This fostered the early development of the independent, mercantile and professional classes – folk of the 'middle sort'. The English king, priests, barons, judges and even so-called 'peasants' may have fought one another for power from time to time but, happily, none ever gained absolute supremacy. British tolerance stems from an individualistic confidence that nobody has the right to infringe the freedom of another. It is toleration born of a refusal to tolerate intolerance.

The British never behaved more aggressively or individualistically than during the Industrial Revolution, but they were not *more inventive* than other nations. Rather, their inventiveness was differently directed. Everywhere else, as far as China (which, but for the short aberration of the last 400 years, has always been the world's most sophisticated and powerful civilization), wealth and power resided in the courts of kings, emperors and the nobility. Britain, by contrast, had a relatively modest, domestic ruling class (the Hanoverian dynasty in particular sometimes seems almost suburban, despite having been imported). And so, inventiveness that in Europe went into making priceless, intricate automata to amuse women and children at court, in Britain went into pots and pans and plates and a hundred simple but ingenious contrivances for the ordinary man and his family. These items, shipped by canal and later by rail to growing industrial centres, sold by the penny. Technology, in English hands, was for use.

While France and Switzerland became the home of precision watchmaking and automation, England became the workshop of the world and remained so into the twentieth century. It was as the poet John Masefield said: quinquireme of Nineveh carried ivory, apes and peacocks, sandalwood, cedarwood and sweet white wine. Stately Spanish galleon bore back to the Spanish king its diamonds, emeralds, amethysts, topaz, cinnamon and gold moidores. And the dirty British coaster carried Tyne coal, road rails, pig lead, firewood, ironware and cheap tin trays. On the foundation of such everyday objects, British manufacturing might was built.

The other difference was our fractured religion. In the eighteenth century, in a reaction perhaps to the excesses of seventeenth-century zeal, Christianity had withered. Hand in hand with mechanical inventiveness went another conviction, born not so much in the established church (which, like so many other eighteenth-century institutions, had emerged a corrupt and complacent wreck, propping up the landed squirearchy) but in the many dissenting churches, and especially the Unitarian movement, strongest in the industrial centres of northern England. Indeed, if the Nield family ever darkened the doors of a church, perhaps because the Chartists were rained off, it was to the Unitarians of Dob Lane that they went, rather than to William Bowen's Baptists.

However, the Unitarians' basically puritan (and, some suspected, covertly republican) ethic was not that far removed from that of other disestablished churches. All these tended to the belief that, while it should never be an end in itself, material gain was an essential precondition of spiritual improvement. Material decency, which had

until then been denied to the vast majority, was viewed by the Low
Church as a basic requirement of full preparation for the world to
come. This was to be achieved by industry – in both senses. In Sunday
schools up and down the land, the young were told how idle hands
would find the Devil's work to do, and this is what I see in the life of
William Bowen, who outside mine and chapel seems to have had no
life at all. It was a dual but total dedication to self-improvement both
material and spiritual.

The 'British Revolution', like its political contemporaries in France
and America, was a revolution of the common man – the man of
parts, who, if he worshipped at all, did so in plain, whitewashed
chapels. The underdog was biting back, brains and merit subverting
the social order. In the coming new world, birth would no longer be
the sole factor dictating privilege. As Beethoven famously wrote to
Prince Karl Lichnowsky (an enthusiastic early patron of the com-
poser): 'What you are, you are by accident of birth; what I am, I am
by myself. There are, and will be, a thousand princes; there is only one
Beethoven!' (Noble individualist – would he have been so confident if
all his paternal ancestors had been called Ludwig?)

As Louis the Sixteenth's head fell from his body on 21 January
1793, one French nobleman, the Comte de Bournon, clutching an
unpublished mineralogical treatise, escaped to London. De Bournon,
a true savant, was one of the few of the European nobility who could
legitimately claim a place in this emerging, meritocratic world order.
The need to publish that book would lead, indirectly, to the setting up
of the world's first learned society for Earth science, the Geological
Society of London (1807).

William Smith, who was never a Fellow of that august body, had
also never suffered the stultifying effect of a proper late eighteenth-
century education. He asked plain questions and sought useful
answers, and it was his intellectual descendants who would come to

dominate the scientific world – men such as Thomas Henry Huxley, Darwin's 'bulldog', who married for love rather than money and strove all his life to make 'scientist' a paying profession. Raised up through the geological bounty of energy, raw materials, transportation and a billion pieces of cheap creamware, the new men of the industrial age revelled in being masters of their own fate. They lifted themselves up by their bootstraps and eventually bought out the mansions of the great with the penny profits of a barge's cargo.

The misfit Raspe was fitted perfectly for this coming world. Lowborn, brilliant, mercurial, opportunistic, he was a wide boy – whom I could portray as a forerunner of my own profession, which, since my brief sojourn in the oil business, has been that of science communicator. Great scientists often need their amanuenses; Darwin needed Huxley, Hutton needed Playfair. As well as a geologist, linguist, analytical chemist and antiquary, Raspe was also a considerable writer, and *The Surprising Adventures of Baron Munchausen* was not the only fruit of his pen. While still in Germany he wrote a number of verse plays, most of which were entertaining enough to enjoy contemporary success. In his last, a chivalric tale called *Hermin und Gunilde* (1766), he even invented a new literary genre – the High Romantic ballad – which was not only successful in itself, but was later taken up by better writers to greater effect.

In Germany, the form was adopted by Gottfried August Bürger (the man who translated *Munchausen* into German) while, on this side of the Channel, it became the preserve of that highest of Romantic wizards, Sir Walter Scott. Scott would unwittingly return the favour he owed to Raspe in his 1816 novel *The Antiquary*, whose action is set against the backdrop of an impoverished Scottish estate and the attempts of its embattled owner to establish lead mines that he hopes will bail him out.

In this early nineteenth-century version of *Downton Abbey*, Scott's plot hinges on a matter of geology. The text refers obliquely to James Hutton, in Edinburgh, and then introduces the very first fictional (and highly unflattering) portrayal of a geological prospector – the 'tramping philosopher', 'High German landlouper' and 'impudent, fraudulent, mendacious quack', Herman Dousterswivel. This pen portrait clearly draws upon a lasting impression that Raspe had made during a journey he undertook through Scotland from 1787 to 1789.

Raspe was described during his lifetime as *ein glücklicher finder*, 'a fortunate finder', who stumbled upon things – precious stones, intellectual nuggets buried in the works of others – and who then, by remounting them newly cut and polished, would repackage them for a wider public (and in the process claim as much credit as he dared). Through his talent for spotting a good story and communicating it, Raspe eventually unearthed some even longer-overlooked ideas about earthquakes and volcanoes and the nature of the Earth's crust. These had originally been published by English natural philosopher Robert Hooke (1635–1703), one of the most sadly neglected yet original scientists of his, or indeed any other, time.

In 1760, while still a clerk at the Hanoverian royal library, Raspe came across a work written by Hooke in 1668 and published posthumously in 1705, entitled *Lectures and Discourses of Earthquakes and Subterraneous Eruptions*. Europe at that time was still reeling from the intellectual aftershocks of the 1755 Lisbon Earthquake, which, as well as dealing a heavy blow to the notion of a benign universe, also contributed hugely to the development of geological ideas. Raspe became convinced that Hooke's overlooked work contained the best explanation, not only of how volcanoes and earthquakes were related, but also of how islands and continents originated.

Raspe decided to re-present Hooke's system for contemporary readers. In fact, it is clear from correspondence that Raspe intended his re-presentation of Hooke as only the first instalment of a larger 'system of the Earth', which he unfortunately never completed. His book, published in 1763, was elaborately titled: *An introduction to the Natural History of the Terrestrial Sphere, principally concerning new Islands born from the Sea and Hooke's Hypothesis of the Earth on the Origin of Mountains and Petrified Bodies to be further established from Accurate Descriptions and Observations*. It was grovellingly dedicated to the Earl of Macclesfield, 'most illustrious President of the most distinguished and most grand Royal Society of London . . .'. The royal took the bait and elected Raspe a Fellow in 1769. Six years later, Raspe (by then in England, and in disgrace) was thrown out, one of very few ever to have their Fellowship revoked. But so diminished had the Royal Society become that Raspe felt the rebuff more lightly than present-day academics tend to imagine. For by then, nearly all *real* intellectual and industrial progress in Britain was being made far out-side the old established structures. Ties to the ancien régime hardly mattered any more.

Sadly, Raspe's attempt to proselytize the neglected work of Hooke suffered much the same fate as Hooke's original, and sank into almost complete obscurity. But Raspe had been right to try. In seeking to explain how fossils had been elevated to the topmost peaks of mountains like the Alps, changes must have taken place in the relations between sea and land since the world was created – changes, Hooke said, that were due to the eruption of subterranean fire which, through what he termed 'earthquakes' (a word he used in a slightly wider sense than we do today), could lift and lower the Earth's crust.

At the time Hooke wrote it, this was a wonderfully advanced piece of geological thinking. It was still ahead of the times a century later. But theorizers lack experience; Raspe's personal experience of mines in the Harz Mountains would have given him an instinctive feeling for the heat that rises continually from the underworld. Perhaps this goes

some way to explain why he was able to make the leap of understanding necessary to grasp Hooke's portrayal of the Earth's surface
as the victim of our planet's internal, heat-driven motions.

At the heart of the Neptunist–Volcanist dispute lay the question
of the degree to which Earth movements, mountain ranges and crystalline rocks such as granite and basalt owed their origin to the
Earth's hot interior. Admitting a 'hot Earth' was crucial, because a
planet that was hot inside would have surface processes that could go
on operating for as long as the heat source lasted, and at least over
the sort of immensely long timescales for which natural philosophers
like James Hutton were calling. This was the time dimension in play;
a hot Earth could also be a very old Earth, whose geological record
offered, as Hutton put it, 'no vestige of beginning, no prospect of an
end'.

Hutton's idea of an eternal Earth was an overstatement compared
to what we now know, namely that our world had a beginning and
will one day die with the Sun. But a lifespan that covers the best part
of ten billion years, only half of which has so far elapsed, comes
pretty close. In effect, Hutton was right. The key battleground on
which this great debate was joined, and eventually won by the
Volcanists, was the Basalt Controversy.

Basalt is the most abundant rock on the face of the Earth – it forms
the floors of ocean basins. It originates as lava, a fact finally put
beyond question by French geologist Nicolas Desmarest (1725–1815)
who proved it while mapping the extinct volcanoes of Puy-de-Dôme
and reported it to the French Academy of Sciences in 1765. Although
his ideas had to wait until 1771 for full publication, his discovery
(immediately notorious for containing the horrifying idea that *la belle
France* had once been peppered with horrid volcanoes) found its
way into Denis Diderot's vast *Encyclopédie* much earlier, in 1768.
Desmarest also concluded that hexagonal columnar basalts –

famously seen at Giant's Causeway in Northern Ireland and Fingal's
Cave on Staffa, Scotland – were also volcanic in origin and not, as the
Neptunists claimed, a precipitate from the ocean. In the Auvergne, he
had found thick basalts that had ponded into lakes of molten lava.
Here, because they were thick, they had cooled slowly and formed
hexagonal columns. He traced these basalts up the very slopes down
which they had streamed from their volcanic vent, and these thin
layers displayed incontrovertible evidence of former flow.

Who should unearth this tiny gem of information, buried deep in
the *Encyclopédie*, but the fortunate finder Raspe himself, at Kassel.
Finding a local angle for the story like any good reporter, he connected the observation on columnar basalt to outcrops of the same
material around the town. In 1769, after consulting Goethe about it,
he wrote it up. His paper was published only a little later than
Desmarest's much delayed original publication.

Goethe himself had been exercised by the Basalt Controversy, and
despite adhering to the opposing Neptunist camp, nevertheless called
Raspe's contribution 'a milestone of German science'. Volcanist ideas
put the poet into a torment that surfaced in these lines, spoken by the
Romantic wanderer staring wistfully across the Atlantic to a simple
paradise:

Amerika, du hast es besser
Als unser Kontinent, der alte,
Hast keine verfallenen Schlösser
Und keine Basalte.

'Better America thy lot, than our old Continent's, unburdened by
ruins of castles – and no basalt!' (He should have written 'no *known*
basalt', of course – some of the world's greatest basalt provinces can
be found in America.)

Columnar basalt's apparent 'crystallinity' lay at the heart of the controversy, though the near-perfect joints are created by the even, contractional stress field that becomes established within a slowly cooling lava mass. Neptunists mistakenly took these crystal-like columns to indicate precipitation from solution. Raspe's great service to science was finding and introducing this new notion to Germany. Sadly, his one *original* idea in the piece (a kind of compromise position – perhaps the basalt prisms formed when basalt erupted on the sea floor?) was mistaken.

Raspe left his adopted city of London for the last time in 1793. Earlier in the year he had visited Aberystwyth on Cardigan Bay, inspected the mines of Cardiganshire and been horrified by the working conditions, which he felt approached 'that of the most forsaken and oppressed Highlanders'. He had made a flying business visit to Cornwall, and then in January found himself back in the capital. Louis the Sixteenth lost his head that same month.

Raspe stayed until the autumn, but wanderlust struck again and he was soon setting off for Haverfordwest, on his way to Dublin. He landed in Ireland in November 1793, describing the 'outrun noblemen' of Ireland who surrendered their mineral rights to English speculators with 'wild notions on gambling success that it grieves me to see how they are misled'.

Seeking to profit by offering somewhat better advice to local landowners, he made his last stop – the last of his life, as it turned out – at an abandoned copper mine at Muckross, near Killarney, on the Herbert family estate. There, after only a few months, he succumbed to a 'spotted fever' (probably what we would call scarlet fever). Tradition has it that Raspe was buried in the graveyard of the estate chapel, near an almost non-existent place called Killegy Lower, in County Kerry.

That church had been long ruined when I visited the area back in the 1980s, while researching an article on Raspe for *New Scientist*. Its roof and megalithic walls were a tangle of creeper, sitting in the midst of a mixed copse on a small hillock overlooking the Muckross Valley. Its presence was betrayed to the traveller on the Muckross Road (part of what the brochures have since dubbed the 'Ring of Kerry' route) only by the massive Celtic wheel cross of the Herbert memorial. Although several older and more modest monuments littered the aged, overgrown graveyard, there was none to mark the last resting place of Rudolf Erich Raspe. The legend was apparently true. The fortunate finder has been lost for ever.

Yet I see him still. The near indestructibility of boundaries in the landscape is nowhere more evident than in London – or any great, prosperous city where prices are high and where, at no time, has possession ever constituted less than nine-tenths of the law. The men appointed to oversee rebuilding after the Great Fire of 1666 (Christopher Wren, John Evelyn and Robert Hooke) discovered this for themselves, when all the grandiose rebuilding schemes put forward by the high-minded foundered on the twin rocks of royal penury and every landlord's right to claim back what had been his, and resume trading. The old medieval building lines were laboriously redrawn and fossilized in fresh new stone. They remain with us to this very day.

The countryside west of the old fire-ravaged City was a different matter. In the fields of Mayfair, old highways such as Fleet Street, Strand, Oxford Street and Piccadilly ran, as they do today, east to west on level ground that is the remnant of a terrace left by a River Thames in ages past, its alluvial sediments concealing bones of the hippopotamuses and alligators that lived and died there during an interglacial period much warmer than ours.

Although this part of London has since become dominated by Victorian and Edwardian buildings, much from the eighteenth century remains – especially north of Piccadilly, in the rectilinear grid of streets that first spread over the fields during that century, the odd winding road, like Marylebone High Street, hints at boundaries of greater antiquity; in that case, the course of a brook now long culverted below ground.

I look out for Raspe on those dark afternoons in late November when, after a gloomy day, the sun dips beneath cloud to blaze briefly down Piccadilly before disappearing below the horizon somewhere beyond Hyde Park Corner, burnishing the buses' rain-blistered backs as they struggle down to Piccadilly Circus. The lights are on; the sky is clearing. Early stars are out. I wait at the kerb to cross Sackville Street, among shoppers and pre-Christmas revellers, noting the geologists walking the other way for some evening meeting at the Geological Society, just as generation after generation has done since the Society moved there in 1874.

The light fades quickly and – as it always does – brings the forgotten landscape to life. Behind me, the grand frontage of Burlington House, with its tall archway leading to the courtyard, vanishes in the mind's eye and is replaced by a curtain wall protecting the grounds of Old Burlington House, town house of the Cavendish family. Their original seventeenth-century mansion has been recently Palladianized in the latest fashion, and its fine, fresh Portland Stone shines warm and yellowish in the evening sun.

Then, turning out of Air Street and scurrying twitchily towards me comes Rudolf Erich Raspe. Though he does not know it, he has just left his lodgings there for the final time. He is, as usual, on the fringes of solvency, hatching a new scheme, hoping for something to turn up, about to chase fugitive metals across Ireland as far as the shores of the Atlantic. He pauses on the opposite side of the junction, leaps the ruts, swerves around a dung pile, sweeps past and is gone.

In less than a century's time, the Cavendishes will have to give up their great town house to pay death duties. The curtain wall, gates and stables will come down and New Burlington House will rise. The fine arts, in the guise of Sir Joshua Reynolds' Royal Academy, will arrive first, occupying the old mansion. Then, in the newly built east and west wings, the Royal Society, the Linneans, astronomers and antiquaries will come, one by one, until 1874 when the Geological Society finally takes possession. By this time, geologists walk in much greater numbers along this pavement. The world they inhabit is industrial and commercial. They are part of a new 'ocracy' of professionals and merchants, whose privilege has not come with land (or even quarrying on that land, as the Cavendishes had, and still do) but with cash accrued by guile and labour.

Raspe's feet, so eager to break free, remained stuck in the eighteenth century's thick clay. But his English patrons, emergent heroes of a new Romantic age, like Boulton and Watt, established a new world that Raspe never lived to enjoy. Driven by industrial need, knowledge about the Earth had poured from holes in the ground as fast as steam, coal or iron ore. Raspe's generation was among the last to find itself trapped within the numbing certainties of estate-based rural life, whose ways had come down, gentrified about the edges but otherwise unchanged, from the Norman Conquest. The next generation, bringing with it men like William Smith and others of the ditching tribe, won their freedom and delivered it in spades to their sons. Enriched by the proceeds of industry and commerce, they were freed to become gentrified themselves, to join their grand London clubs and publish learned papers about the discoveries they made during their forays into the underlands.

There seemed, in that time, no limit to industrial growth. The world was vast, its natural powers invincible, its existence eternal. Men, on the other hand, were few and puny. It was their duty to wrest what they could from nature's grip and give it to their sons and daughters to make better lives for themselves. But soon – perhaps not soon enough – new generations of geologists would realize that in their well-intentioned rapacity men were already outstripping nature's forces. They would come to recognize that in coming ages, humans would need to rein back their demands on the bounty of past times and learn to live within the planet's present means.

This discovery came as a shock, and many are still having difficulty mentally adjusting to it, let alone acting accordingly. Geologists in the late nineteenth and early twentieth centuries had discovered that the all-important time dimension was almost limitless. This finding only served to reinforce the idea that feeble humans were powerless to alter the Earth's timeless equilibrium.

Even many climate scientists fail to acknowledge the importance of historical evidence. They persist in believing that the past can have nothing useful to tell us because what humans are doing to the climate is, they mistakenly assume, unprecedented. It is not so. The Earth has already endured climate change every bit as sudden as that which industrial activity is bringing down upon our heads.

Geologists have already emerged from their holes in the ground with a prophecy and are foretelling that a catastrophe that happened 55 million years ago, long before humans evolved, might (this time thanks to us) be about to happen again. Nature has done it before, without our help.

At first this idea seemed absurd, but we now know otherwise. And it all happened as a result of deep-seated magmatic processes within the heat engine of the planet, whose geological significance Rudolf Erich Raspe was among the first to grasp.

6

Timefathers

That our sons may grow up as the young plants, and that our daughters may be as the polished corners of the temple.

Book of Common Prayer, 1662

Most of Britain's great northern industrial cities lie somewhere near coalfields, though not that many lie on them. Typical coalfield topography is less than ideal for urban development. The best practical situation for any city is a relatively level site, with a short supply line to its source of power. Manchester, my father's birthplace and the world's first truly industrial and commercial city, gets this about right.

Manchester grew at great speed from its small, ancient origins, and did so according to the dictates of its emergent industries. Manchester was not a port until men made it so, and although its nineteenth-century textile industries were rooted in the wool trade of the sixteenth century, today you will look almost in vain for medieval or older buildings (there are in fact three listed at Grade One). A Saxon church, dedicated to St Mary and recorded in the Domesday Book, once stood on the site of Manchester Cathedral, now hemmed in between a bend in the River Irwell, Victoria Station and the Arndale Shopping Centre. Its Late Perpendicular heart, built after the demolition of the older church in 1422, is surrounded and overprinted by Victorian restorations and enlargements. The nave of the cathedral is broader than it is long, largely thanks to the addition of so many private chantry chapels, which hint at a strong mercantile history long before the Industrial Revolution. St Mary's was then a rich parish church, the fine carved misericords under the seats of its choir stalls depicting rural scenes (including one of a hunter and his dog being cooked by a hare).

To build their church, Manchester's medieval masons chiefly used the local Collyhurst Sandstone, though after the church was promoted to the rank of cathedral in 1847, Bath Stone made an appearance in the new chapter house. Many repairs were made after Manchester's 1940 blitz; but it is odd that, medieval masonry and Domesday Book pedigree notwithstanding, and despite the best efforts of Shambles Square's timbered buildings, Manchester Cathedral's drab context makes it look like just another piece of Victorian Gothic Revival. The cathedral precinct of Britain's pre-eminent industrial city is hardly a rival for that of York (though, as a semi-Lancastrian, it pains me slightly to admit it). Manchester's true architectural glories, such as its university, belong firmly to the nineteenth century.

Coal-measure rocks stretch in an arc around the heart of Manchester, from Warrington and Worsley in the north-west, through Wigan, Bolton, Bury and Oldham to Macclesfield in the south-east. Most of the city itself lies on a plain, underlain by younger rocks of the Permian and Triassic Systems, deposited 250 million years ago by ephemeral torrents running in sediment-choked wadis, which crossed the searing deserts of the supercontinent Pangaea. But the early cotton mills had no need of coal. They were powered by water in the hills surrounding the plain. It was the import of raw cotton through Liverpool, and the conversion of mills from water to steam, that drove Manchester's growth. New transport links consolidated its position. The Mersey and

Irwell Navigation made those waterways passable for cargo vessels. The Bridgewater Canal, with its wondrous tunnels and aqueducts (and whose opening, in 1761, is generally taken as the starting point of the Industrial Revolution), carried the third Duke of Bridgewater's coal from Worsley to feed the mills' voracious steam engines.

The new industrial economy demanded a new workforce, which it sucked off the land. My English ancestors at Risk Farm outside Mossley were, by the 1820s, working in the mills of Oldham and Failsworth and living in the terraces south of the Oldham Road. As it developed, this new technical economy needed another army of people; people equipped with new knowledge – *useful* knowledge.

Established ancient seats of learning offered an education irrelevant to these needs, and so through the middle years of the nineteenth century the great civic universities grew up in response. Their buildings reflect their times and mirror their industrial context. Manchester's Victoria University (as it used to call itself) sits like a great railway terminus on the Oxford Road. To construct this monument to civic pride and useful knowledge at the zenith of the city's economic confidence and power in 1895, the city fathers turned to Alfred Waterhouse, designer of London's Natural History Museum. The job of overseeing its construction would be passed down from father to son. Few trades run in families quite like architecture.

Unlike Manchester's cathedral builders, Waterhouse used Carboniferous sandstones, nearly all of which travelled a modest forty miles from Darley Dale, in Derbyshire. Very similar in appearance to the Forest of Dean Stone used in Cardiff Castle and William Bowen's box tomb, this is a fine-grained, buff-coloured freestone; and today the building, having been recently cleaned, has returned to this original colour. But in the late 1930s, when my father first walked through its great entrance archway, a few decades' exposure to Manchester's vile air had caused the Waterhouses' Gothic fantasy to take on a mourning veil. In a nightmarish smut black, its aspiring pinnacles and steep, fish-scale roofs vanished upwards into eternal and dully glowing smog.

After leaving school in 1937, my father cycled each day from home in Didsbury. The Nields by then had made the leap from blue collar to white, and the Ted Nield who had placed my father on the rock in Llandudno had left his father's house in Failsworth (where sister Lotte still worked at the mill), moved to the leafy south-west and bought a Rover. His son, on leaving school, loved nothing more than music and science and so, being advised that 'people like us go in on the ground floor and work our way up', he entered the University's Department of Chemistry. Here, at the recommendation of Didsbury Central School's Principal Teacher, Mr F. K. Maddrell BA ('a very efficient boy'), he had been taken on by Dr Albert Edward 'Max' Gillam, lecturer in organic chemistry, as departmental spectroscopy technician.

University departments then were not the chummy collegiate places they became in less formal times, and I get the feeling that (with the honourable exceptions of his boss Gillam and one Michael Polanyi, whom he also liked and greatly respected) my father regarded the academic staff – one or two future Nobel Prize-winners among them – as rather one-eyed and philistine. They would have been surprised, no doubt, to discover that the lad in the brown coat with the *Odyssey* in his pocket found them wanting.

In between long spells in the laboratories below the Oxford Road pavement performing spectroscopic analyses for staff and students, Father teamed up with some technical buddies from Physics, forsook the common room and spiced up the lunch hour by exploring the secret byways of Waterhouse's vast fortress of useful knowledge. Until one day they found their way into the attic behind the long-silent clock, perched on the gable end of the John Owens Building, overlooking the quadrangle. Taking pity on its seized and web-hung mechanism, the technicians started doing what came naturally.

On a typically sodden autumn evening some weeks later, lights were on all along Oxford Road, each in its smothering ball of fog. The University's serried windows glowed through its sickly yellow. In the quadrangle, beyond the archway, wet, ripple-marked Haslingden flagstones (just like those used to pave London's Trafalgar Square) gleamed. No one that day had noticed that the University clock had been silently telling the correct time since the first tea break, for nobody believed it possible. Towards the end of the working day, a person with keen hearing passing below the Owens Building might have heard the muffled whirr of a mechanism gathering strength. And then, shockingly loud in the quiet away from the road's noise, the clock struck five – for the first time in longer than anyone could remember. My father put on his bicycle clips for the four-mile ride back to Parrs Wood Road, anticipating somewhat naively how he and his friends would emerge from obscurity next day, be thanked for their unpaid work and take a bow.

Alas, the sudden and uncomfortable intrusion of time into the business of inculcating eternal verities proved less than welcome to the University authorities. In a memo to all departments, the musically named vice chancellor, John Sebastian Bach Stopford (himself the son of a mining engineer from Wigan), demanded to know *who mended the clock* in tones that suggested the culprit would be unwise to expect much gratitude. Happily, like most management witchhunts, Stopford's failed – and Father's debut as a guerrilla repairer never came to light.

Nobody confessed, no one was exposed; the clock was officially restopped and the affair soon forgotten. Had father been unmasked, he might never have received the official reference from Max Gillam for his war job, as spectroscopist for Swansea's Magnesium Metal Co. This would have been serious – at least for me, for that move enabled him to meet William Bowen's granddaughter, on a blind date in 1941. History, narrative of the time dimension, turns on such contingencies.

Those five unexpected bongs from the University clock on that afternoon in 1939 fell on the ears of another recent recruit to the establishment, a man destined to become better known than any of his contemporaries, great or small. Then a junior lecturer in neighbouring Physics, Dr Bernard Lovell, later Sir Bernard Lovell FRS (1913–2012), became known chiefly as the creator of Jodrell Bank, the world's first large steerable radio telescope. This made him one of Britain's foremost public scientists, and his affable but authoritative presence became a frequent and welcome feature of radio and television in the 1950s and 1960s. Lovell also quit the University for war-related work in 1939 (in his case, at the Telecommunications Research Establishment in Dorset). In due course, his son Bryan became a geologist, and it was in his company that I emerged, one bright day in October 2012, from a graffiti-covered concrete underpass below a dual carriageway somewhere in Hertfordshire.

Barely two months before, Bryan had delivered the eulogy for his father, a task that awaited me in another six. We walked slowly. Although Ulysses is not that strength which in old days moved Earth and heaven, he remains strong in will. We spoke about fathers, the passage of time and the accidents of history that flowed from the Manchester of the late 1930s, down to our first meeting at the Geological Society.

Bryan, who had lately come to the end of his term as the Society's President, was leading me to some rather special holes in the ground. For one thing, they were holes in the ground in the South-East of England, which made them special anyway. These, however, so long abandoned by men and wasted by time, were also a precious archaeological site. Unlike at Horn Park, here I would find no barbed-wire fences or spiked steel palisades. Archaeologists, like farmers, prefer the defences offered by secrecy and camouflage; and for this reason, I had already promised to keep details of the precise location to myself.

The first people to notice – and bemoan – the vanishing of our holes in the ground were geologists, who relied upon them for research and teaching. But the problem varies according to geology itself, and land prices. In most areas of Wales, Scotland and northern and western England, rock is old and hard. Holes cut into it tend to degrade slowly. The landscape around Haslingden and Rossendale, for example, will bear the beautiful scars of flagstone quarrying at Lee, Cragg, Moss and Landgate, Thurns Head, Ab Top and Musbury Heights for many centuries. The land is high, agriculturally poor, good for walking and wind farming but not much else. Population is relatively sparse, and land prices correspondingly low. On the other hand, holes dug in rocks that are young and soft, and where the low-lying land they support is rich, sought-after and expensive, disappear quickly.

So it is hardly surprising that it was a geologist from the Thames Valley who first raised the alarm. Professor John François Potter began doing so almost as early as Alec Clifton-Taylor, in 1968. At that

Professor John François Potter points out the ferruginous gravel blocks
in the chancel wall of St Mary's Ripley.

time, for those of us outside the South-East it was still 'business as usual', and in 1974, when I began studying geology at university, academics estimated that a quarter of all university-level geological field instruction took place in quarries.

Geologists were not the only group to benefit from the existence of small local holes. By the 1990s, of all designated Sites of Special Scientific Interest in South-East England, well over half (and almost eighty per cent of SSSIs in East Anglia) were mineral extraction sites, either ancient or modern. Both ecologically and didactically, 'holes in the ground', as Potter wrote in a letter published in 1995, constitute 'man-made environmental assets' that should 'figure high as items for preservation'. Significantly, he had been driven to write those words in outraged response to reading a fieldwork teaching aid, which to his horror had posed students the question: 'Would this quarry make a good waste-disposal site?'

As Bryan and I walked slowly up the rising path before us, I aired the subject of disappearing quarries and told him about a recent journey I had made with a certain Professor Potter. 'Ah – would that be "Graveyard" Potter'?' Bryan asked. Geology is at once a big subject and a small world.

The rocks that Bryan was taking me to see are evidence of truly stupendous climate change that happened 55 million years ago, entirely as a result of natural processes, and which enables us to picture the dire consequences for us if we humans do not change our ways. But how much more dramatic was this event than, say, the climate changes of the recent past? The world's climate has never been stable, and its fluctuations during the last million years or so – the period known broadly as the Ice Age – have not been exactly trivial.

I had wanted to make two journeys to help me understand the differences between what one might call 'normal climate change', induced by complex but understandable changes in the Earth's orbit, and 'catastrophic climate change' induced by forced changes in the composition of the atmosphere. As it happened, both involved looking at rocks in what is now the Thames Valley – rocks that, despite being of widely different ages, are both composed of pebbles, cemented together. My visit to Hertfordshire with Bryan was the second of these journeys. As we slowly advanced towards our goal, I recounted my day with John Potter.

One huge pit is not very useful to a scientist wishing to recreate lost worlds and understand the geological story that lies concealed in the landscape. What geologists like best is many small holes, all over the countryside, providing the widest possible sample. When John Potter was growing up, as an evacuee in rural Surrey, surrounded by sand and gravel pits, nobody would have dreamt that one day soon nearly all these would be gone, and geologists would have to rely for exposures on the last remaining accessible samples, which had been accidentally preserved one thousand years before by Anglo-Saxon church builders, desperately scrapping around their stoneless native soil in search of something solid to build with.

In South-East England, ancient churches have become artificial outcrops. These buildings, growing directly from their local soil, can tell us about much more than their own brief history, fascinating though that is. As old pits degrade, fill and become housing estates, their fabric has become the last surviving memory of the underland, and its tale of past worlds that lie locked away beneath our feet.

In Hertfordshire, Bryan and I were walking on the northern perimeter of the Thames Valley, whose soft rocks and superficial deposits recount the last 100 million years or so of Earth history. The detail preserved in that record becomes ever greater the closer we approach the present, and much of that recent geological history has to do with the changing Ice Age climate, its effect on Old Father Thames and the much older tributary rivers that join from the south as they flow off the Weald. It was a story first worked out by geologists studying the gravels and sands that once lay exposed in tens of thousands of local workings – now mostly vanished.

We are used to seeing the tame rivers of today's landscape and forget how, before our ancestors canalized, embanked and controlled them with levees, weirs and sluices, they were chaotic, braided torrents choked with bars and islands of gravel and sand, and running willy-nilly across wide floodplains. As these wild old rivers flowed and switched back and forth, each would leave behind a telltale ribbon of gravel to document the history of its life course. These gravels are all that remains of them, and over a century of careful analysis has gradually taught geologists the full truth about how the rivers of southern England responded to repeated advances and retreats of ice.

Step onto a suburban garden patio deck on a dewy morning, and you may find it covered in a hundred glistening trails, each made by one specimen of *Helix aspersa*, the common garden snail. If you could catch every specimen now hiding in the grass, you could sample its DNA and match it to the trail it had made the previous night and so unravel the history of the pattern beneath your feet – which snail went where, which came first, and so on. The gravels of the south of England are like this. Each gravel trail, left by a vanished river, has its own genetic fingerprint, reflecting the unique geology of that stream's headwaters.

By carefully identifying and counting the different pebbles, geologists can show that by about 780,000 years ago, the son of 'Great-Grandfather Thames' had migrated south, passing north of Harlow to join with a greatly extended River Medway (with its load of distinctively different pebbles sourced on the Weald of Kent). The Thames then reached the sea between the modern ports of Ipswich and Harwich – where the Rivers Orwell and Stour now run. So even

'Grandfather Thames' did not flow through the site that would one day be London.

What eventually pushed the river's next 'generation' further south was a great climate shift within the Ice Age known as the Anglian Cold Stage – a global cooling event that began about 450,000 years ago, pushing the ice sheets further south than ever before or since. From our perspective, this was the 'last but two' glaciation, and it was the most severe. Ice reached a kilometre or more thick even in Hertfordshire, shortly before the earliest humans to occupy Britain left artefacts and skull fragments at Barnfield Pit, Swanscombe, in the gravels laid down by the Thames during the warm period that followed this cold snap.

Swanscombe lies south of the modern Thames, from which you will gather that the Anglian Glaciation had overwhelmed the river's old northerly course and forced it to flow along the route it has followed ever since. As the ice melted, oceans filled and sea levels rose – on that occasion, higher than today, so that the rivers that formed immediately afterwards had a topographically higher 'base' level (sea level, to which rivers erode) than modern ones. Today's Thames, responding to the current lower sea level, has therefore cut down into the deposits of its old valley, leaving the gravels deposited by its former self clinging to the margins, on terraces.

By studying such gravels, geologists have come to know that the Thames's southerly tributaries, such as the Medway in the east or the Rivers Mole and Wey in the west, are much older. Before the Anglian Glaciation they flowed over counties far north of the present Thames, even reaching as far as Hertfordshire. They were foreshortened by glacial bulldozing, but the gravel trails they left behind (for not all were completely swept away by the Anglian ice), exposed in a thousand gravel pits, have allowed us to trace their ancient routes through their lost territory.

As we walked, I told Bryan how, just a few days earlier, I had met John Potter in Surrey, in a village called Send, still 'remote' despite lying just beyond the M25. We met at its church, a place I had chosen because of a rather peculiar coincidence in Potter's curious story.

As field geologists tend to be, John is remarkably fit and busy for his years. He was born in his grandmother's house in Putney, London, in 1932, and soon moved with his parents into a new house in Isleworth – built, as it happened, on the site of a former brick pit. There is a certain irony here, considering how much effort John has expended, as an educator, journal editor and writer, in alerting people to the way holes in the British landscape are being lost beneath suburban sprawl and general tidiness. It shows that, while we think of it as a recent phenomenon, reusing quarried land, especially in the overcrowded South-East, is an old tradition. It's just that these days, for every ten holes that are filled, barely one is opened.

John clearly has a genetic predisposition towards holes in the ground. His engineer grandfather caught gold fever while delivering an engine to the delightfully named Flyshit Creek, near briny Lake Annean in the goldfields of Western Australia. Nothing of the local township of Nannine remains, though it once boasted a population of 30,000. John still keeps a biscuit tin containing his grandfather's rock samples, and a ring made from a nugget of alluvial gold that he found. John's father carried on in engineering, though as a teacher; and so at the outbreak of war found himself entitled to the privilege of evacuation en famille, along with the sixty or so children for whom he was responsible. It was this event that probably made John Potter a geologist.

John remembers standing with his gas mask, coat and rucksack at Clapham Junction, amid the throng and confusion, waiting to join the train that would take them all to the supposed safety of the English countryside. He remembers anticipating eating the sandwiches he carried and the great disappointment when, before having the chance, barely twenty minutes after squeezing aboard, they were disembarked at Woking. A bus took them the few remaining miles and brought them – to Send.

I arrived first and had just retraced this very same journey. It was a soft spring day in May, overcast but bright. The announcement of water restrictions, following three years of exceptionally low rainfall, had heralded almost a month of near-continuous rain (in fact, 2012 turned out to be England's wettest year on record). As the lane became more countrified, its gutters ran with water under the ferns. The water table was close to the surface.

The final mile to the Church of St Mary the Virgin led along an even narrower lane from a crossroads so deep and green that it was hard to believe how close I was to London's suburbs and its ring of motorway. But for birdsong and a distant cockerel, all was silent. The fresh spring leaves in their apple green moved in the breeze, but were not yet stiff enough to rustle. The air was heavy with hawthorn flower and the yellow blooms of a broom-like hedgerow I couldn't identify. The entire width of the dead-end track, signed by a fingerpost to the church only, now ran with spring water.

John's description of this junction when he first saw it in 1939 could have come straight out of a Samuel Palmer painting: deepest England, safe, ancient, mystical. Huge chestnut trees arched over the unmetalled roads; and beyond the hedgerows, along the terrace above and stretching forever in every direction, lay endless gravel and sand pits. John would spend his boyhood years exploring these local quarries, developing the impression that, whatever direction his life would take, it would have to involve the outdoors.

Send Church stands within a hundred yards of the River Wey, on a slight eminence, surrounded by a graveyard. I entered through a blackened oak lychgate. A church was recorded here in William the

First's Domesday Book of 1086; the original building would have
been Anglo-Saxon. How much Anglo-Saxon construction remains in
the present structure it is not easy to say, though John Potter's work
applying geological knowledge to help untangle the complex history
of buildings is radically re-evaluating the amount of original Anglo-
Saxon work that still survives.

Builders of these first churches in southern England faced one
great problem – the almost complete lack of decent building stone.
Transport by any means other than water was impossible, so they
either contented themselves with wood (the Anglo-Saxon Church of
St Andrew, in Greensted-juxta-Ongar, Essex, being probably the
oldest partially wooden building in Europe), or else made do with
whatever lay beneath. The only building materials occurring anywhere
near to Send were the river gravels that lay just under its topsoil.

The slight rise on which the church stands, slightly higher than the
modern floodplain, consists of the Kempton Park Gravel, dating
from the last Ice Age. About six metres thick, it is a mixture of gravels, sand, silt, clay and peat, and was deposited by the ancestral Wey –
which has since cut down through its old deposit leaving it high
(though not very dry). Importantly, the water table – the level beneath
which the pore spaces in the gravel are filled with water – is here very
close to surface, as I had observed.

The church's masonry betrays the fact that it was made out of anything and everything that came to hand. There are flints, blocks of
chalk, pieces of tile and roughly shaped boulders of not very resistant sandstone, many of which have been repaired and replaced by
cement. The quoin stones, at the corners of the building, show signs
of greater selectivity, being blocks of either Chalk rock (some containing flints and fossils of the sea urchin *Micraster*) or the older
Greensand, which is not very resistant to weathering and has often
been replaced. And dotted among these brought-in materials are

rough, shapeless lumps of rusty rock, the colour of burnt ginger-bread, containing ill-assorted pebbles cemented together in a finer matrix.

Chalk rock and Greensand outcrops can be found within two or three miles of the site, which, being close to the river, was probably able to receive these stones by boat. Flints, which make up the bulk of the wall, are ubiquitous on the Chalk, so would also have been gathered locally. But those rusty-red conglomerates of pebbles had come the least distance of all. This stone is the church's most native component – the Kempton Park Gravel itself, formed on this very site up to about 70,000 years ago, when Surrey was a chilly, periglacial wasteland, criss-crossed by torrential outwash rivers and their fans of poorly sorted alluvium. This was the material I had come to see.

The entire geological history of the region could be traced in these walls – from the time of the Lower Greensand, about 120 million years ago, through the age of the Chalk (to 65 million years ago, the end of the Cretaceous and the extinctions that also carried off the dinosaurs). Youngest of all were these gingerbread gravel blocks from the Ice Age. But there was more to it than that – and more to this ancient building than a condensed field guide to local geology. Its very fabric embodied the Earth's immense antiquity; it contained the remnants of organisms – some extinct, like *Micraster*, some still living like the *Viviparus* freshwater snails seen fossilized in the Wealden age 'Sussex Marble' – that told the story of life's evolution.

There was poetry in the thought of how the Anglo-Saxon builders had founded their church a thousand years ago upon these mute witnesses that contained clues to more knowledge about the Earth's history than the creed to which the builders cleaved; a creed that for so long denied the Earth's and human antiquity. I also reflected that more local rock lay exposed in this church's ancient walls than can now be seen in situ for miles in any direction.

Startled by my arrival at the collapsing boundary wall of the cemetery, a roe deer sprang suddenly away, breaking my line of thought. She threw herself into the swollen Wey, swam to the opposite bank and bounded off through the long grass of the water meadows, leaving a dark trail in the dew. Behind me, a car pulled up by the lychgate.

I got to know John Potter when he wrote an article for the magazine I edit, *Geoscientist*. It told of how he had first become interested in the geology of churches, a subject now given the dignified disciplinary moniker of 'Ecclesiastical Geology', which he hoped the article would help introduce to the wider geological world. The story of his own involvement in this new subject began in 1975 when he was asked by the British Geological Survey to help answer an inquiry from a member of the public concerning a church in neighbouring Ripley, three miles from Send. The inquirer asserted that the church appeared to be 'made of gravel'.

On visiting the church, which sits at an acute angle to Ripley High Street, Potter was able to confirm that this was so. The oldest part of the mostly nineteenth-century church is its chancel, whose walls are strikingly different from the neatly knapped flints of the nave, with its yellow dressings of Jurassic oolitic limestone so typical of Victorian work. The chancel, however, included not only flint nodules but large, rusty-brown blocks of pebbly conglomerate, laid as neatly as the material allowed, in rough courses.

The terrace on which Ripley Church sits is the so-called Taplow Terrace, whose gravels were deposited maybe 100,000 years ago. In 1975, the source of the brown blocks in the walls was immediately made obvious to Potter, because next door to the church lay a gravel pit (now covered by houses). To a geologist, the oddest thing about these gravel blocks is the fact that they are cemented by iron; gravels this young are not usually cemented at all, which is why they are quarried for aggregate. But Potter had noticed that within these quarries, odd, irregularly shaped blocks of cemented gravel were often seen, abandoned because they were of no use.

The process by which the blocks had become cemented is a continuous and continuing one, and occurs at the water table itself. Here, where the pore spaces are partly filled with water and partly with air, iron precipitates out of solution, gluing the pebbles and sand grains together. This cement is not very strong at first, just enough to hold the fabric of the rock together as it is hewn from the layer in which it occurs. But once exposed to the elements, the coating of the pore spaces becomes converted into hydrated iron oxide – better known as rust.

As the blocks dry out, they become harder, eventually forming a useful local building material in an area where none is available except by means of long and arduous transport. Small wonder then that, having observed this process, perhaps while digging graves, Anglo-Saxon builders seized upon the ferruginous gravel blocks, not just for their durability but also because they had the immense advantage of being on site.

From Send, John and I made the three-mile drive to Ripley. From the crossroads at Church Lane, the road climbed onto the Taplow Terrace. As we gained the level ground that was once a river flood-plain, he told me, we were seeing former gravel diggings either side of the road. These former pits now look like little more than slightly hollow fields; others have been filled and built over. In the few that have been allowed simply to return to nature, their hummocky micro-landscapes remain visible beneath dense huddles of trees.

These gravels were the witnesses to what 'normal' long-term climate change, typical of the Ice Age in which we live, can achieve. It can change sea levels by over a hundred metres, bring the edge of the Northern ice cap as far south as the Thames Valley, and change the courses of major rivers for ever. This variability is the nature of the planet we live on. Only our tiny time sample, our human brevity, our failure to grasp the time dimension at the scale of the Earth herself, fool us into thinking that we can expect climate to remain constant. The message of the rocks in the vanishing gravel pits of the South-East is that the mere orbital wobbles of the planet over tens and hundreds of thousands of years have done this, and will continue to do this, all on their own.

Not a man to do things by halves, during the thirty years and more since he discovered how early church builders used local ferruginous gravels, John Potter has visited no fewer than 10,000 churches and ruins in the Thames catchment, checking their masonry and plotting the ancient ways of the Thames and its tributaries. As part of this process, he has tried to bring some scientific rigour to the rock descriptions bandied about by archaeologists and architectural historians, many of whom have used obsolete geological terms – including the term 'puddingstone' – for Potter's favourite ferruginous gravel.

This is a particularly unfortunate terminological gaffe to make in Hertfordshire, where another distinctive rock, the Hertfordshire Puddingstone, crops out. This remarkable rock, noted for its hardness, was also widely used for church building, and it was to a natural outcrop of it that Bryan Lovell had led me. Fields on either side of our rising path had been ploughed to corduroy, ready to receive their next crop. The rich brown soil was noticeably stony, with beautifully rounded black and brown pebbles of flint lying everywhere.

Bryan, who has something of the Jedi Master about him, waved one of his two long walking sticks in their direction and told me to ignore them. These were not the rocks we were looking for. In fact, they were the remains of river gravels dating back a mere million years or so. 'That's all Potter-type stuff,' he joked. Eventually we reached the contour of a small copse, which crowned a ridge in the field to our left. 'Strike directly across the field,' Bryan said, pointing. 'You'll find the quarries inside the wood.'

Bryan was wise not to trust himself on the freshly turned soil, which sank spongily beneath my boots. I began to notice, scattered among the pebbles and the ploughed-in straw, several larger, more angular cobbles – mostly with freshly broken surfaces. What attracted my attention was the smoothness of these fracture faces; for these cobbles were of a rock that was itself composed of smaller pebbles, set in a very fine greyish matrix.

The fracture faces were unusual because, unlike Potter's gravels, which broke *around* their constituents like a fruitcake, these cut straight across both pebble and matrix equally, leaving a smooth surface on which round outlines could be traced – by eye, but not at all by touch. The matrix between the pebbles was just as hard, in other words, as the pebbles themselves – which tells a geologist that it is likely to be of the same mineral. The pebbles were of flint, eroded from the chalk bedrock 55 million years ago and reincorporated in this younger deposit. Then, they had been cemented together by chemically similar and equally obdurate silica. The rock they made was the Hertfordshire Puddingstone.

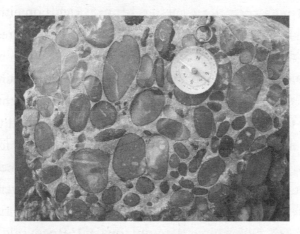

Hertfordshire Puddingstone – a fossil shoreline deposit from 55 million years ago, when a natural global warming catastrophe reached its peak.

'Puddingstone' is an old-fashioned geological term for any rock containing rounded pebbles – what modern geologists would call a 'conglomerate' – the old name being a picturesque reference to the texture of a pudding full of large, juicy fruit. Herein lay the confusion of terms that so annoyed John Potter. His gravels could once have been described as puddingstones, written with a small 'p'. But since that usage has become old-fashioned, the term is only ever seen today with a capital 'P', signifying one particular puddingstone, of a certain age, from a certain place. In this region of England, Hertfordshire Puddingstone is the most abundant durable building material around, and many a handsome church has been built of it.

Earlier in the day, Bryan and I had visited the village of Standon and its parish church of St Mary. Outside the churchyard, beneath an oak tree on a patch of green, we examined a huge monolith of the county's most famous rock. A single nodule of silica-cemented conglomerate, it stands about three metres tall on its modern plinth (dating from 1904), and viewed from one angle in particular it bears a strong resemblance to a female with prominent buttocks and breasts. As a notice beside it says, the stone is traditionally known as the 'Breeding Stone' and was formerly incorporated into the wall of the churchyard.

The term 'breeding stone' was once applied to all Puddingstones, not just those blocks that happened to look like fertility goddesses. It referred to the notion that these stones were alive, and grew spontaneously in the soil. Tradition holds that the Standon Breeding Stone was at one time an object of pagan veneration, and it may be so; though it is also possible that it was simply a boundary marker, which might explain its former position in the churchyard wall.

Whatever its archaeological significance, the Standon Breeding Stone has survived many thousands of years' exposure to the British weather – testimony to its enduring mineralogy. It was the same recalcitrant rock-type that the Romans had been seeking in the secret quarries towards which Bryan now directed me. To win this tough material, they had opened the workings where the copse now stands, its incredible hardness being the key to its economic importance. The Romans needed it, not to build with, but to make querns on which to grind corn – then as now, the staple of life.

The outcrop, just below the soil, was also the reason why an unproductive copse occupied the middle of this otherwise rich field, its plough ridges splitting around it like a river around a boulder. Generations of farmers had discovered to their cost that any plough attempting to break ground there would shatter on the reef below. Lying exactly where it was deposited 55 million years ago, this patch of Puddingstone outcrop constitutes the tattered, fossilized remains of a pebbly shoreline that once fringed a tropically hot sea.

Fifty-five million years ago the Earth was very different, with global sea levels several metres higher than they are today. If you had been able to look down from space upon what would one day become Britain, you would recognize, like an unfinished sculpture half embedded in its marble block, an approximate but familiar outline, lying about ten degrees south of where our northward-drifting islands sit today. To the east lay the early North Sea; to the west, the Rockall Trough.

On its north-eastern seaboard, your eye might pick out a wide embayment with a very straight, east-to-west aligned southern coast – the distinctive maw of the Moray Firth. The sharp south-westward switchback in the coastline at Rattray Head north of Aberdeen already forms its distinctive wedge shape. South of where Edinburgh now stands, the coast swings east and south, following roughly the present shoreline from East Lothian as far south as Norfolk. There, the hot, shining, silver coastline, with its rounded flint pebbles and silica sands, turns south-west, slicing through what is now Hertford-shire, on its way towards Devon.

At that time, the crust underlying this entire area of Western Europe was in tension. Deep cracks had caused a volcanic province to develop over what is now Skye, Islay and Northern Ireland. Huge basalt lava fields had built up, whose eroded remains can still be seen in the Cuillins, and the cliffs of the Hebrides and Antrim (including the polygonally jointed columnar basalt lavas of Giant's Causeway and Fingal's Cave).

But, while Scotland and Ireland were being buoyed up by this rising magma, to the east the North Sea Basin was dropping, like the keystone of a splaying arch. Sand, eroded from the rising land between the North Sea and the Atlantic, rushed in, transported by torrential rivers that formed huge deltas where they met the ocean – massive sediment piles perched precariously on the edge of the deep-ening rift. The supply of this sediment was not constant, however. Pulses of the buoyant magma, rising from deep within the planet (just as Rudolf Erich Raspe imagined) periodically jacked the land up another notch, rejuvenating the processes of erosion.

The great delta fans, destabilized by earthquakes, periodically foundered downslope to deeper water – their mud and sand sliding into the abyss as turbid, bottom-hugging avalanches of remobilized sediment. These 'turbidity currents', as they are called, spread out into the basin, smothering the sea floor with sand layers that became progressively thinner, further from source. One day, the pore spaces in these thick accumulations of deep-water sand would receive a hydro-carbon *pneuma* emanating from organic-rich rocks deep below them in the sedimentary pile, cooked up under heat and pressure by the internal heat of the planet, and so over long ages become reservoirs for North Sea oil.

Meanwhile, out west in the widening Atlantic, sediments were also accumulating – only much more slowly. Far from the reach of turbidity currents, these deposits consisted of nothing more than a diaphanous mixture of windborne dust and the shells of microscopic sea creatures that floated in the waters above and rained down constantly as they perished. Since these microscopic shells were secreted by living things, their composition reflected the chemistry of the water in which they grew, and for that reason a number of chemical markers can be used today to determine the climatic conditions prevailing tens of millions of years ago. Such markers are known in the trade as 'climate proxies'.

Deep-sea drilling has recovered many thousands of kilometres of core from the floors of the world's ocean deeps; thin leaves of sediment that have built up, year after year, recording the Earth's climate in layers that can be numbered with amazing accuracy down to mere thousands of years. This might not sound very accurate when a few thousand years is enough to take humans into 'prehistory'; but to a geologist, dealing habitually with the time dimension at a planetary pace covering thousands of millions of years, it constitutes a fine-tooth comb. The cores show that something highly unusual and dramatic suddenly happened to our planet's climate 55 million years ago (roughly 50 million years before our humanoid ancestors first made their evolutionary appearance).

One of these chemical proxies – the ratio of calcium to magnesium in the shells of marine organisms – allows us to find out the temperature of the ocean-bottom waters at the time of deposition. Plots of such palaeo-temperatures show that, 55 million years ago, the oceans warmed by between four and five degrees Celsius. Given that the planet was already very warm, as far as we know the Earth has never become hotter in the last few hundred million years. Because the date of this climate event occurs at a boundary between two geological periods (the Paleocene and the Eocene), the temperature spike has been called the 'Paleocene-Eocene Thermal Maximum', or PETM.

Most of us are now aware that one of the main controls of average global temperature is the amount of carbon dioxide in the air. Carbon dioxide allows the Earth's atmosphere to trap more of the sun's heat – the well-known 'greenhouse effect'. It is not the only gas to do this, nor is it the most effective; but it is the most abundant, and small variations in amount can have disproportionately large effects. The amount of global carbon residing in the atmosphere in gas form is strongly affected by geological processes and, as in many other ways, human beings are now the most significant geological process of all.

Carbon is all around us; in the air, the sea, in living things, in sediments, and it circulates between these reservoirs in a slow geochemical cycle not unlike the (much faster) one that regulates the interchange of water between airborne vapour, oceans, rivers, lakes and ice. However, the 'carbon cycle' is slow enough to include rocks. Just as there is only a finite amount of water on the planet's surface, only so much carbon is available to circulate around the carbon cycle. It follows that the more carbon that is 'held up' in one reservoir, the less there must be in others. For example, the amount of carbon absorbed by plants during photosynthesis reduces the amount in the atmosphere. If that 'fixed' carbon is then deposited in rocks, in the form of coal or limestone, or as oil, gas, peat, lignite or methane hydrate, it becomes locked away, or 'sequestered'.

Limestone is important in this cycle because much carbon becomes locked away in the shells of marine organisms, which are mostly made of calcium carbonate. Organisms get their carbon from seawater, where it arrives in solution after being eroded off rocks exposed on land by rainwater made acid by the carbon dioxide in the atmosphere. The rain corrodes the rocks, minerals dissolve and the carbonate salts that result from that reaction are washed into the sea, to be absorbed by living things.

The balance between all these processes is dynamic, and can change radically through time. During the Carboniferous Period, for example, a lot of organic carbon ended up sequestered in the form of limestone during the Early Carboniferous, and as coal during the Late. Abetted by the fact that tough, newly evolved plant materials called lignins were then still indigestible to the agents of decay (fungi, bacteria), the Carboniferous world saw massive increases in atmospheric oxygen and record lows of carbon dioxide, as plants proliferated but did not fully rot. For this reason, despite its popular reputation as a time of steaming jungles, the Carboniferous Period was actually a global 'icehouse', with very little global warming and extensive glaciations in higher latitudes.

The period in which we are living today is also an 'icehouse' time, with low sea levels and ice caps at both poles. While orbital wobbling controls the size of those ice caps, the Earth itself influences its surface temperature through two major factors: the presence of a landmass at the South Pole, encircled and kept permanently cold by a circumpolar ocean current and covered in ice many tens of kilometres thick, and the elevation of the Himalayas. These mountains, and the vast Tibetan Plateau beyond them, together with other young active mountain ranges on Earth today, expose huge amounts of rock to intense erosion, and thus draw more and more carbon dioxide out of the air, cooling us down.

Icehouse periods like this are rare in Earth history. About twenty per cent of the Earth's past has been spent this way, as opposed to about eighty per cent in the greenhouse (climate tending to 'flip' between the one stable state and the other). Moreover, during icehouse worlds, the proportion of time spent in the relatively less cold periods in-between the advances of the ice caps (which we tend to call glaciations) is also low, again, merely twenty per cent. Our current interglacial climate – having lasted, roughly, for the last 10,000 years – is therefore typical of only about five per cent of Earth's total climate history. Twentieth-century doubts about Hutton and Lyell's doctrine of Uniformitarianism were indeed justified. Human experience is simply not at all representative of the grand sweep of Earth history. It is not even enough to help us appreciate the true nature of 'business as usual'.

One crucial deep-sea sediment core, which includes the PETM horizon at 55 million years before present, was taken from the western North Atlantic back in 1997. Chemical analysis of the carbon in these sediments went a long way towards explaining why that sudden hike in global temperature took place.

Many chemical elements can have more than one possible atomic weight, depending upon the number of neutrons (which have mass, but no charge) present in the nucleus. These atomic varieties are known as isotopes, and their slight differences in weight mean that they may become separated out during natural processes. Rainwater, for example, contains relatively more H_2O molecules containing atomically light oxygen, because rain forms by the condensation of moisture that got into the air by evaporation. Evaporation, for obvious reasons, preferentially selects lighter atoms.

Processes in the bodies of living things favour, and tend to concentrate, carbon's lighter isotope, carbon-12. This means that 'fossil' carbon, in limestone, coal or oil, for example, by virtue of deriving from once living matter, is isotopically light. Carbon dioxide released by burning those materials therefore also carries this same 'light' signature. In the crucial core from 55 million years ago, scientists discovered two alarming facts. One was that, at that time, excessive carbon was injected into the atmosphere over a very short period indeed, geologically speaking – a mere 10,000 years. This is the same amount of time that has elapsed since the end of the last Ice Age heralded our species' cultural development into the Earth's most potent geological agent. The second alarming fact was that the isotopic ratio of this excess carbon was 'light'. That could mean only one thing. The carbon that caused the runaway global warming at the Paleocene-Eocene Thermal Maximum, 55 million years ago, had come from fossil sources.

The injection of carbon dioxide into the atmosphere was both massive and geologically sudden. Global temperatures shot up. Water close to the ocean floor rose from an already very high eleven degrees Celsius (this was already a 'greenhouse' world) to fifteen degrees Celsius. The world of that time had no more land ice left to melt; but still the oceans rose because they expanded, as all things do when they get warm. And so sea level, already high, rose by between five and six metres more – creating beaches in what is now Hertfordshire.

Whatever caused this catastrophic event did not last, but Earth's climate did not regain its equilibrium for over 100,000 years – which we can take to represent the time required for natural erosion and sequestration processes to swab all that excess carbon dioxide out of the atmosphere.

The super-resistant siliceous rocks I was hoping to see in the Roman quarries of Hertfordshire had been deposited at the height of this climate catastrophe, a fact that Bryan Lovell has been tireless in pointing out. Silica is a mineral whose solubility in water is acutely affected by temperature. And so the final link in this historical chain of causation is forged: the unequalled global temperatures of 55 million years ago also explain how the Hertfordshire Puddingstone came by its obdurate cement.

I gained the copse's edge and ducked beneath its bare, overhanging boughs. Against the light beyond the trees, a roe deer sauntered away with exaggerated discretion. The wood's floor was covered in fallen leaves – hardly ideal conditions for hunting Puddingstone. I climbed a little way in. Sure enough, the ground had been worked, for among the lanky ash trees lay a number of shallow craters from which slabs of precious Puddingstone had once been pulled free.

Archaeologists have recognized an ancient roadway leading down-hill from these quarries, crossing the field and heading straight for Ermine Street, father of the A10 dual carriageway that parallels it today. Quarried boulders would have been rolled down this declivity and then carried to Bishop's Stortford, to be fashioned into querns; though some may have been made on site too. It is impossible to gen-eralize because the quarries were worked more or less continually until, finally, nature was left to reclaim them in perhaps the sixteenth or seventeenth century.

I tried to envisage the scene, imagining the holes deeper, the banks higher, no trees; an exposed and rather uninviting place when the east wind blew, or rain came out of the west – though at least the ground would have drained well enough. An explosion at my feet brought me back to the present – a hare, no longer trusting its camouflage, lost its nerve and bolted, zigzagging away, kicking up the leaves. Shafts of sunlight shone through light mist between the trees into the irregular egg box of leaf-deep hollows. A forgotten industrial landscape had rarely seemed so forlorn.

The stone that was dug here by Romans (and, maybe, even remoter ancestors than they) was as essential to their agrarian lifestyle as coal was to William Bowen's world, and as oil is to ours. Their economy may have been what we tend to call 'pre-industrial', but without querns to grind grain, there would have been no bread. Quern-making requires hard, dense rock, and in a world largely lacking cheap long-distance transport, it must have been enormously valuable in this part of England. I felt close to where extractive industry, and the rise in atmospheric carbon-dioxide levels that it has created, began.

I started back along the furrows to where Bryan was waiting on the path, chatting to a passing dog-walker who probably took him for the farmer he resembled in his waterproofs and cap. Like his father before him, Bryan is a remarkable scientist. He studied at Oxford and Harvard before lecturing at Edinburgh University in sedimentology, the science of how sediments accumulate. His special subject was the origin of deep-water sand bodies, and his arrival in Scotland just as North Sea oil was opening up was another happy accident of history. The sandstones that formed, pulse by pulse in the subsiding trough of the North Sea, were just his kind of stuff; and they were then being targeted by oil companies from all over the world as potential reservoirs for the black gold.

Precisely because oil is first found in the mind – 'seen' for the first time because somebody has believed it possible – an exploration geologist has to be able to make predictions (you could call them educated guesses) about the best places to drill. He or she will make those calls upon which hundreds of millions of dollars may then be wagered. Such guesses rely on information from drilling records and seismic sections, combined with an understanding of how sedimentary basins work. They also depend on knowing what kinds of geological structures might trap any fugitive hydrocarbon molecules, cooked up from organic-rich source rocks deeper below and percolating upwards like the *pneuma* once inhaled by the Delphic oracle. Bryan's special knowledge of deep-water sands and how they accumulate enabled him to interpret samples taken from expensively drilled cores, and to predict the compass direction in which other potential reservoir rocks would exist.

Of course, in reality even the most perfectly set trap can fail to catch anything, and the most propitious-looking geology might produce nothing but strong brines, or the oilman's rueful 'puff of dust'. Oil may be found by the mind's eye, but you can only prove that it is really there by drilling, and nine out of every ten holes drilled produce nothing more useful than further geological information.

Bryan left academe in the 1980s and worked at BP Exploration for fifteen years, where he became a legendary player in this sort of high-stakes game. He has now returned to academe, as a Senior Research Fellow at Cambridge University, and studies the way that ups and downs of the Earth's crust, induced by magmatic movements below, affect the way sediments accumulate at surface. This work has led him to look beyond the deep-water sediments in which North Sea oil was found, to those deposited at the same time along shorelines – the beach sediments that became the Hertfordshire Puddingstone. And as a man who has done more than most to help us liberate the power stored in the carbon reservoirs of the past and bring it into the present, he has been struck by the stark message that those rocks carry to us across a gulf of 55 million years.

As I walked back towards him, my boot hit a stone in the spongy soil. At my feet lay a typical Puddingstone chunk, recently fractured, probably by the ploughshare it destroyed. I wanted to carry home a

Hertfordshire field, near a Roman quarry. Above a certain contour, where the Puddingstone rests upon the underlying chalk, boulders and cobbles surface. The block to the right has been recently fractured. The smaller, rounded cobble to the left is a hammer-stone, used for shaping blocks on site.

sample of this intriguing rock as a souvenir; but this one was a
boulder – too big. What I was after was a cobble. (Geologists, of
course, assign precise measurements to these vague colloquial terms.)
Lying next to it, however, was another specimen of just the right size.
I picked it up – noting immediately how easily it sat there, as though
made to do so.

Like any tool sculpted to fit the hand, there was only one way to hold
it. My right thumb curled around a protuberance on the stone's 'upper'
side (there is no other way to describe it), while my fingers wrapped
comfortably about its rounded body, so presenting to the ground an
almost perfectly hemispherical surface, rather like a large, broad pestle.
Without pausing to think further, I gathered it up and brought it to
Bryan. He too enjoyed the way it seemed to leap to the hand. 'You've
made a find there,' he said. 'I think that's a hammer-stone.'

It had seemed like a tool because it was one. Someone, during the
first 1,500 years or so of the last two millennia, had once used this
Puddingstone mallet to shape a larger block. Being the most durable
substance for miles, in the absence of expensive metal tools the only
object capable of making a dent in Puddingstone is Puddingstone
itself. Gradually, blow by blow, this hammer would have smoothed
and rounded the surfaces of a saddle or beehive quern, destined to
become a precious possession, the central object in some humble
dwelling, maybe passed from generation to generation, like my brass
candlesticks, bearing the gouge marks made, I imagine, by my great-
grandmother in murky Nixonville.

No hammer-stone could last for ever. In the end, it would either
be broken in use or discarded, like this one, when its task was done.
This may well have been the tool of just one craftsman, made for a
job at hand by someone living at the very beginnings of industry
which has, generation by generation, improved our wealth and quality of life beyond the mere subsistence afforded by the meagre present.

The tool he had made and used lay under the meadow possibly for centuries before being jostled by plough, harrow and frost, year after year, the last surviving memorial of its maker, who fashioned and wielded it at a time from which, but for the gentle shapes of the Hertfordshire hills, almost everything else has been lost. Tool of a trade, it now sits on the mantel at home, next to a grimy leather-cased tape measure, the very one William Bowen once carried underground at Merthyr Vale to mark off the stalls for each gang of colliers to work.

Bryan's career in oil began after he had returned from Harvard to Edinburgh. He was, as he has written, 'a tyro ... characteristically proud of his fledgling specialist abilities' – in his case, the knack of distinguishing one type of sandstone from another and diagnosing the environment in which it was deposited. He soon met BP's chief geologist, Dr David Jenkins, who had come to Edinburgh to talk about the exciting new developments offshore. They talked over dinner about the Forties Field, discovered just two years earlier in 1970. Not then 'brought in', Forties was to prove the North Sea's biggest field, 110 miles east-north-east of Aberdeen, containing five billion barrels of oil.

BP geologists were still debating whether the sands of the Forties Formation, in whose pore spaces the oil had been trapped, had formed in shallow water or deep. Jenkins already had an inkling that the answer was 'deep', but kept it close to his chest. A few months later, in January 1975, the year in which production from Forties began, Bryan found himself tearing lids from core boxes containing samples of the precious pay dirt. He saw a sandstone that, as he put it, was 'almost too perfect to reveal its origin': so perfect, in fact, that it appeared to lack the little telltale clues a sedimentologist needs to diagnose sedimentary environments.

But at last, Bryan found what he needed. The rocks, he determined, had formed relatively close to the source of the sand, but in quite deep water, as turbidity currents rolled down the ancient continental slope. From this Bryan predicted that there would probably be more potential reservoir sandstones in acreage lying to the south-east of Forties, and which might therefore be worth exploring (which they were – the Nelson Field was brought in, by Enterprise Oil, in 1988).

A decade and a half after that discovery, the story of the Forties Formation developed further, as he listened to researchers from Cambridge University speaking at BP's offices in Aberdeen. They were explaining that, according to their research, just as the Forties sands were being laid down, Scotland was undergoing repeated phases of uplift, as hot magma rose up periodically beneath it.

At this moment, like a clock suddenly and unexpectedly chiming in the gloom, the time dimension intruded itself. Everyone knew that the sandy accumulations making up the Forties Formation had formed episodically. Between the beds of sand, muddy layers marked times when normal deep-sea sedimentation – the gentle rain from heaven – returned. Mud like this eventually formed the seal preventing the oil that migrated into the porous sandstones from leaking away to surface.

At this point, two pieces of knowledge – pulses of sediment accumulation, constrained in time by accurate fossil dating, and periodic magma intrusion – meshed together. The episodic inrushes of sand had been caused by periodic uplift of the land, which rejuvenated erosion. These periodic uplifts had been the result of intruding magma pulses beneath the volcanoes of North-West Scotland and Northern Ireland. Research has since shown that the peak phase of uplift came, in fact, in two sub-phases. One raised the Earth's surface west of Scotland by at least 490 metres, and another (about a million years or so later) caused 300 metres of uplift further to the east.

With the very exact dating that had by then become possible, it turned out that these two pulses of uplift precisely bracketed a date in Earth history that will by now have become familiar: 55 million years ago, when glittering white beaches snaked through Hertfordshire, and when sediments way out in the North Atlantic recorded one of the fastest and most dramatic climate-warming events that the Earth has ever experienced. Bryan believes all this is more than coincidence. But what *in nature*, triggered by uplift, could have caused a catastrophic release of isotopically light fossil carbon at such a rate and scale?

Around the Earth's ocean shelves, today as then, large amounts of methane are produced by the partial decay of organic matter. Under the right low-temperature conditions, this becomes trapped in a cage of water molecules to form a gelatinous substance known as methane hydrate. Vast quantities – thousands of billions of tonnes – of carbon can build up in this metastable form. And because hydrates are so volatile, the methane they sequester can be released very quickly if, for example, sediment is disturbed by submarine slides, pressure is released or temperature raised. In other words, simply making large areas of hydrate-laden ocean floor warmer and shallower, or causing them to slump into subsiding deep-water troughs, would be a sure-fire way of suddenly releasing large amounts of methane, most potent of the greenhouse gases, into the atmosphere.

Our atmosphere is thought to contain about 3.5 billion tonnes of carbon as methane. But at any one time, the methane hydrates accumulated on the cold shelf edges of the world's oceans contain many *thousands of billions* of tonnes – an amount comparable to all the Earth's coal resources combined. Release, even of a part of this carbon reservoir on such a scale and over a mere thousand or so years, while not catastrophic in the Hollywood disaster-movie sense, would certainly be sudden enough to overwhelm the carbon cycle and cause runaway global warming.

However, finding out where the excess carbon of 55 million years ago came from is academic bow-tying. From a practical point of view, locating its precise source is a side-issue. We already know, from the incontrovertible evidence of the carbon isotopes, that that source was fossil carbon – of some sort. That is as much science as we need for 'policy purposes'. What we are doing to the atmosphere now, nature has done before. And the proof, as Bryan has written, lies in the Puddingstone.

During the PETM, as much as one thousand billion tonnes of carbon were dumped in the atmosphere in a single millennium. Human beings have so far been responsible for between a third or half as much – between 300 and 500 billion tonnes. All of that carbon did not accumulate just during the two centuries since the Earl of Bridgewater's canal first carried coals to Manchester in 1761; humans began burning carbon long before the late eighteenth century. But it is worth remembering that rates change dramatically through time, so although we started much earlier, it is a startling fact that roughly *half* of all that excess carbon has been placed in the Earth's atmosphere since not 1760, but *1970*.

The Puddingstone is telling us that unless we want to see Hertford-shire at sea level again, this is a 'climate experiment' we should not be repeating. Yet there is little sign, despite all we know, and all the summit conferences, communiqués and agreements, that we will achieve a slowdown in carbon emissions very soon. If more people listened to the rocks, learnt the parable of the Puddingstone, realized that our so-called 'climate experiment' has already been run and the Earth has seen it before; if they appreciated how *very* inconvenient it would be should all that happen again, scientific advice based on computer models (which can more easily be waved away as fantasy) might be met with less scepticism and more action.

I confess to feeling a reckless, childlike excitement about the impending disaster of catastrophic global warming; the vanishing island states and the drowned coasts where most people on Earth live. Presented with a row of dominoes set up to fall, the temptation to push the first one over is almost irresistible. But mature judgement, the sense of responsibility that knowledge brings and, above all, concern for future generations, teach us to resist such temptations.

When I first saw the spoil heaps of Aberfan in 1968, and imagined the excitement of standing on Tip No. 5 and perhaps running or falling headlong down its white-streaked black slopes, I was experiencing that same impulse. Were it simply a matter of their dramatic presence on the mountainside, I would have to admit that I regret that the coalfield's old black shale heaps have now become part of a forgotten landscape, together with all physical traces of the mine that made, and finally destroyed, the little Welsh township where my mother was born and where her ashes have now returned.

And yet, they had to go. Those piles of shale and tailings, perched high above everyone's heads, rich enough in the combustible black element actually to catch fire, were not just an unacceptable reminder of the tragedy that engulfed Pant Glas School. They posed a continuing danger because they had been sited irresponsibly, blind to the geology of a mountain that spelt trouble for anyone with eyes to look and the prepared mind to see; the Pennant hillside, built from sandstones shot through with impermeable shales that bring treacherous rivers to surface, bleeding, night and day, into those vast heaps' underbellies.

Because geology spells trouble, we must heed what the underlands are telling us. If we listen to them, understand their Delphic utterances, they will give us warning. An immeasurably bigger pile of waste carbon now sits above our global village. From what the Hertfordshire Puddingstone tells us, we know it is more than just a worry. It spells certain disaster, if not for us, then for the children of those who put it there. But it remains invisible, and that is the main problem.

Carbon as carbon is highly visible. We used to see it everywhere, in the grime of Manchester and Aberfan and all towns built on coal. We think we have cleaned up our act since then. We wash the mourning veils from the facades of grand Victorian buildings, raze the waste heaps, repaint our windowsills, see the sheave wheels broken and the pitheads toppled and valley turned green again. We think the environment is the better for it. We may not have work, but at least we can hang out the washing.

Despite the leisure, cycleways, recreation areas and urban nature reserves, all we have succeeded in doing is treating a few local health hazards. Meanwhile, the carbon pile that we *should* worry about mounts. We deal with the tiny things we can see with our eyes, instead of the whole we can only conceive of in our minds. We fiddle with small-scale tidiness while the fires burn on. When we most need to keep faith with our past, listen to what it is telling us and act so as to save our posterity, our focus narrows on our little backyard.

The possession of knowledge imposes a duty to act. That's the trouble with it. The rocks have spoken. Geologists have, in a hundred thousand holes in the ground, invoked the spectres of the past and they have prophesied. Geology also tells us how we can use the Earth sustainably, on a scale in both time and space that the planet can absorb without harm. It warns us about what will happen if we fail. Do we want future generations to curse us for making *them* pay off the debt with which we have saddled them with our wanton waste of energy, dragging Earth materials halfway across the planet and back? If we could return from the grave, would we rather be thanked warmly for our good work, or face a witch-hunt? Can we live in the knowledge that the future might not revere us for what we did, and keep our monuments bright? For if we put its many blessings out of mind, we forgo the past's greatest gift – the warnings that it brings to us out of time.

Epilogue

Happy Valley

On the day that Fabienne and I drove to Aberfan for our noon
appointment with the man from Bereavement Services, the year had
just begun its decline. A chill had entered the wind, like the first ignor-
able symptom of a terminal illness. For the moment, though, the
patient was still full of life. Tall towers of cloud sailed the sky, deposit-
ing torrential showers that fell across the valley walls like long net
curtains blowing from an open window. Their grey, slow-moving veils
thinned the green of the conifer plantations covering the scars of the
hillside like sticking plasters.

Then at midday, the downpour suddenly stopped. The sun shone.
Everything gleamed: the black cemetery roads, the forgotten old
tombs and the remembered new, bright with fresh blooms. We
emerged from the car, I carrying the little square oak casket with its
brass plate and the engraved names of two whose journey was now
done.

William Bowen's restored grey granite shone rather garishly in the
sunshine. To the right of William and Margaret's pedestal memorial,
a carpet of green AstroTurf had temporarily replaced the new granite chippings, and through the hole in its centre I could see the sides
of a square pit, floored with red sand. The official examined the

cremation certificates and put them in his pocket. He took the casket and lowered it in, handing me a fistful of black soil, which I cast in with a farewell thought. It landed square on the nameplate and obscured it for ever.

This was our second farewell ceremony; for my mother and father's mingled ashes had been divided in two. One half, according to my father's wishes, Fabienne and I had already taken to Dorset and scattered over the Inferior Oolite capping a tree-tufted hilltop not very far from Horn Park Quarry, where father and son had spent some of their happiest hours releasing ammonites from their bed of time.

In Aberfan, after handshaking and thanks, we left Mossfords' men to their work and walked back to the car. I looked out towards the featureless desolation of grass where John Nixon's masterpiece, Merthyr Vale Colliery, was once sunk at the cost of so many lives, on the far side of the River Taff beyond where Smyrna once stood. I found myself thinking again about the *Odyssey*.

I own many translations of Homer's epic, most handed down from my father's collection: Chapman, Pope, so strongly contrasting; Butcher and Lang, epic and grand; Shaw (T. E. Lawrence) and W. H. D. Rouse; E. V. Rieu's influential Penguin Classics prose treatment, the first version I ever read, and Robert Fagles's blank verse, which I could never finish. All have their virtues, but the point is there can be no single, definitive rendition of Homer. As Butcher and Lang (my personal favourite) point out in their 1879 preface, all translations are pictures taken from a lost point of view – the point of view afforded by the era in which the translator worked. Each rendition not only reflects the Bronze Age mind of the author, but the minds of those who interpreted it for their contemporaries. Every translation of the poem inhabits its very own lost landscape.

Since so much of the poem is devoted to Odysseus, son of Laertes, we are apt to forget that the work covers three generations of father and son. The relationship between these generations, the veneration of the ancestor and the duty to keep faith across all the divisions that life throws across our paths, haunts it – just as it has haunted human dreams since antiquity.

After sitting with my own father during his final days in a nursing home, the futility of Telemachus' efforts to reconnect with Odysseus strikes me now as one of the epic's most poignant aspects. Poor Telemachus; the noble youth's heart may have been in the right place but until he redeems himself in the end, he is pretty useless.

I felt useless too, as I watched my father struggle for words to describe the dreams that he increasingly mistook for reality and heard him slurrily recount conversations he had had there; conversations in which he and I had spoken with men and women already long dead when I was born. In his waking moments, he would occasionally paw after shades of past things that only he could see. Regularly he would mistake the living for the dead, and at night, so he told me, he would sense the approach of a hand, reaching up to him from the darkness below the bed.

People often remark how the old may fail to remember or recognize their friends and relations. Distressing as it is, I believe it to be a secondary phenomenon. As his dementia took hold, the chronological framework of events and people that once gave structure to his memory fell apart and became mixed together, as though the river of time had suddenly broken its banks and become a great, turbid lake. While this had the effect of mixing his past and present, it was the complete antithesis of my temporal vertigo, which grows and feeds upon knowing the true relations of things in time. Father was not surprised by the juxtaposition of people from different periods of his life, because he now lacked the necessary frame of reference – an internal chronology.

Because the past defines us and tells us who we are, another side-effect of this random mixing was to make my father forget who he was himself. Without that chronological structure, which puts the events of a life into distinct periods and gives them names, correlation became impossible. The identity of those around him, past and present, could no longer be measured by reference to time. The triangulation of his life's landscape collapsed, like an electricity pylon struck by lightning.

His loosening grip on time bore a deep irony. Since those early days under the roof of the Owens Building at Manchester University, Father's fascination with timekeeping machines never abated. When he left his house for the last time, it contained almost fifty – from antique pendulum clocks through to gimcrack plastic digital give-aways (which he could never resist, because he never grasped that thanks to new technology, accurate timekeeping had become a commonplace that we heedless folk now take for granted).

He would complain loudly, spring and autumn, when the change to and back from British Summer Time loomed; but he could never part with any of them. And yet despite all their help, in the end he wandered out at four in the morning on a Sunday night, believing it to be four in the afternoon on the following Tuesday. During his last year, the man who once mended clocks for amusement lost the ability even to read them.

Four days before his ninety-third birthday, in the small hours of the night (and after some unresponsive days, during which I could no longer tell whether he knew or even saw me) my father felt at last the welcome touch of that hand from the icy dark below, and died. By that time, the boy had become father to the man; in our last conversations it was clear that in his eyes, I was now the father. 'It's a long time, isn't it, since you last had to do this for me?' he asked, as I took spoon in hand to help him eat his cornflakes one morning. He knew that a bond of obligation united us. And that debt should naturally run from son to father, towards the past. With his mental clock broken beyond repair, all he had left to judge our relationship was the flow of dependence.

Afterwards, clearing his room, I found among his many illegible notes on scraps of torn paper an unsealed envelope. On it, printed with obvious effort in the shaky hand of an eight-year-old child – the child placed on a rock in Llandudno for a photo – were just four drooping lines, which I think of as his last words to me.

Dear Father
I can't think.
When are you coming to visit me?
I miss you terribly.

In Homer's epic, the generations finally reunite in a cathartic bloodbath – the slaying of Penelope's suitors and their traitorous paramours. In life, ancestors always escape us. Unlike Odysseus, they never return, unless invoked, like the shade of Tiresias in the land of the Cimmerians, by the touch of those persistent objects that tether us together in the flow of time. We are left, in life as in geology, to knit whatever scraps remain into a patchwork representation of what has vanished. Time's erosion leaves us, as it leaves the Earth, with tattered and ambiguous remnants from which to weave a tale that seems to make sense to us now; a snapshot, from a unique viewpoint that will itself soon disappear for ever. Without at least making the attempt to examine the past, nothing around us can have any meaning. We owe our past everything, and if we ask it the right questions, it will answer, and look after us. Otherwise we face the future blind and helpless.

Fabienne and I left Bryntaff Cemetery. High above Merthyr

Mountain, two red kites rode the back of a strengthening wind. Their plaintive, falling calls descended through the cold air, and they vanished as the sun weakened and a new veil of rain drew across. Now that phase of life is done. I sit on my brownfield site, its only (and final) occupant, by a smouldering log among the pillars of antiquity, feeling like some unworthy goatherd. I keep inventing tasks I must do to keep faith with the timefathers.

I do not mean that I want to repeat the things they did. To learn from ancestors, and from the past, is neither to live in the past nor to force it to conform tidily to suit the present. Our ancestors did the best they could for us. (At least, mine did, mostly, and I am lucky in that.) Our job is, like them, to find our own way in our own time, armed if possible with an understanding of why they did what they did. We must revere our ancestors, even for what we now think of as their follies. They only seem that way in hindsight. They did what they must, and so must we; but we have choices that they did not. We are blessed with the ability to understand when rocks speak.

One thing I have felt compelled to do has been to read Anatole France, whose complete works I now own. On the day of the interment I was still working my way through these often brilliant but largely outmoded writings, which had once seemed great in my father's eyes as well as those of the Nobel Prize committee; for they shared the lost perspective of their time. I recalled some lines from *Pierre Nozière*, one of the master's volumes of fictionalized reminiscence that seems fresh still today. The words were spoken not by a living voice, nor yet a dead one, but by stones of the church at St Valéry-sur-Somme. And the stones said:

> Look you, I am old but I am beautiful. My pious children have adorned my robe with towers and steeples, gables and belfries. I teach them to labour; I instruct them in all the arts of peace. My

children grow up within my sheltering arms, and when their tasks are done, they go one by one and sleep at my feet under the grass there, where the sheep graze ... they owe me all; for man is man only because he remembers.

Learn, then, from me that sacred way of hope that saves our land from ruin. Let your thoughts dwell in me so that they may reach out beyond yourselves. Behold this, which your forefathers handed down to their sons. Labour for your children as your ancestors laboured for you. Not one stone of mine but brings you a boon and reminds you of a duty. Look on my cathedral ... and revere the past. But bethink you likewise of the future. Your sons will know what gems you, in your turn, have inwrought upon my robe of stone.

I have also returned to Llandudno, to take one last picture that completes, definitively, the series unwittingly begun in 1928 when my grandfather first placed my father on a limestone boulder, clutching a stick and a rose.

It was a bright, grey day at the end of the summer season. Llandudno's broad streets looked pretty in pinks and pastels, but were largely deserted. Outside the hotels with predictable names – Grand, Victoria, Waverley, Bryn y Mor – old couples with small dogs sat on municipal benches, looking out to sea. A man, too old to be still sporting the DA haircut of his 1950s youth, sang in the doorway of an empty shop. To the accompaniment of his karaoke, he crooned an Edwardian song, 'The Sunshine of Your Smile', in the style of Roy Orbison – neatly conflating two heyday periods of the British seaside.

Oversized Llandudno Station, anaemic and starved of trade, was undergoing a much-needed refurbishment. Leaving it behind me, I made for the eastern promenade and then headed towards the pier,

above which lay Happy Valley, Lord Mostyn's Jubilee gift, up a short
diagonal fork from Marine Drive, a scenic route that circles the Great
Orme from just behind the Grand Hotel. Everything was much as I
remembered; the Jubilee water fountain, the Gorsedd stones used for
the druidic ceremonies of the National Eisteddfod and placed there
just two years before I first visited.

I climbed past the empty cafeteria and up the main drive, until I
found the path leading away to the left. Now that the shrubs and
saplings among which I was photographed in 1985 have become
trees, the old mining gallery to which the path leads is invisible.
Elephant Cave opens today on to a light mixed forest of ash, hazel,
sycamore, hawthorn and black pine. The cave still boasts a few
charred remains of fires on its bare earth floor, but to judge by the
debris lying around them, glue sniffing has been overtaken by cheap
supermarket cider.

Outside in the gulley, and far from having vanished beneath the
soil, as I had feared, the famous boulder looked slightly uplifted. The
reason was not hard to find; the leaf canopy now makes it too dark
for most plants to grow, so underfoot only a little grass sprouts among
the sphagnum and hart's tongue ferns. The luxuriant turf of 1985 has
shrunk almost to nothing, receding from the boulder like gums from
an over-brushed molar. It might no longer stand a proud metre tall as
it did in 1928, but this time I could not miss it.

I had brought prints of the photo series with me, so that I could
triangulate the view and place the camera tripod as close as trees
permitted to the spot where my grandfather had stood eighty-five
years ago. After composing the shot I set the camera's timer, and ran
to plant my boots exactly where my 1985 brogues, my Start-Rite
sandals of 1964 and my father's scuffed shoes of 1928 had once been,
and perhaps still are, simultaneously in this place, further upstream
in the river of time. Nobody – apart from some rackety magpies

cackling in the branches – witnessed this pantomime, as early leaves fell.

In 1887, on 21 June – a day that would one day become my birthday – Victoria's Golden Jubilee celebrations, with their band music and speeches, could have been heard in the distance. Far to the south, William Bowen celebrated his thirty-first year, though not, naturally, with the demon drink. He had been deacon of Smyrna for a decade by then, and two decades a miner. In two years' time, the Glamorgan Technical Instruction Committee would be formed and William would begin his mining course, after his shift was done. In one year's time, the man who would place my father on this rock would be born, high on the Pennines in Mossley, in the house belonging to his grandfather, Edward Nield the stonemason.

I closed my eyes and saw the trees retract into the earth and the stones grow tall as the soil shrank away, revealing dry, grey limestone rubble beneath. The path widened to a road. Carts, iron tyres stinging on the gravel, returned their heavy burdens to the galleries, which filled and blocked. Elephant Cave vanished. The rock of ages flew back to its place high in the cliff from which it had fallen, reunited with the stratum of which it was once part, and had lain for 330 million years since the doomed age of bright coral seas.

Far below, where the pier had stood, the sea withdrew beyond the horizon. Ice covered the Irish Sea, flowing south from Scotland and meeting the glaciers that flowed out from Snowdonia, until the land was lost beneath in its rasping grip.

The Earth's robe of stone stores many bounties and we owe it everything; but there is none richer than its story, which only we, of all Earth's creatures, have evolved the means to read. Everything that can happen has happened in the Earth's long life. The rocks remember all. Let them speak.

Happy Valley, Llandudno, 2013. Author for scale.

Acknowledgements

I owe a considerable debt to Professor John Potter for first introducing me to the subject of Ecclesiastical Geology, and the intimate relationship between the churches of the Thames Valley and their native soil. He was also the first to put into words the growing concern that Earth scientists, professional and amateur, have long felt about the decline in the number of man-made exposures in Britain. Whether the growing trend towards the 'amenitization' of certain sites can entirely or even partially make up for the loss of real, working or until recently working small quarries, I doubt – and so does he. They are a good deal better than nothing. But can one or two 'city farms' make up for the experience of the real thing that I once enjoyed?

Alun Richards, *y Dyn Llechu* (The Slate Man), author of many works on the Welsh slate-mining industry and a neighbour and friend in Swansea for many years, assisted with identifications and was a fount of information on the making of bricks – those man-made rocks for which we share a peculiar enthusiasm.

Sandy Whyte and Hugh Black kindly invited me to Aberdeen and showed me around Rubislaw Quarry, which they had lately purchased and partially drained, and discussed their plans to return it to the people of Aberdeen with a new visitor and conference centre.

Jonathan Pickup of Bright Books drew my attention to texts on Roman mining in Britain, while retired mineworker and historian of the Welsh mining industry Ray Lawrence BSc provided, through his many publications and correspondence, a wealth of information on the Merthyr Vale Colliery, from sinking to closure.

The librarians of the British Geological Survey allowed me access to rare bound copies of the *Stone Trades Journal*, while as usual Wendy Cawthorne (Geological Society Library) listened to lists of things I thought I wanted, and then found the ones I actually needed.

Simon Morgan, managing director of Mossfords Monumental Masons, Cardiff, spoke enthusiastically and candidly about the stone trade, in which he is the third generation in his own family to work. He also placed Mossfords' company records at my disposal and made site visits with me to Aberfan – as well as showing me around the company's extensive stoneworking facilities on the Cowbridge Road during the restoration of William Bowen's tomb.

Dr John Cubitt, one-time managing director of Poroperm Laboratories Ltd, gave me my first taste of oil business consultancy back in 1984, and I must thank also his wife, Professor Cynthia Burek (University of Chester). She, writing an opinion piece for *Geoscientist* many years later, introduced me to the concept of 'rock miles' and the need to think consistently about transport costs, not only for foodstuffs, but for all raw materials.

Sam Scriven (Dorset County Council Earth Heritage Team) showed me around the Horn Park Quarry SSSI, and Bob Chandler was a fount of information about the history and geology of the site. I should also thank Marcus Chambers of the wonderful Beaminster Museum and all the many helpers – especially Jenny Cuthbert and Brian Earl – who made me so welcome when they were obviously very busy. My thanks go also to Dr Jonathan Larwood, Tom Sunderland and Dr Colin Prosser (Natural England) for their help, and for their work protecting the geological heritage of Britain.

Dr Barbra Harvie, University of Edinburgh, kindly took me on a guided tour of the oil-shale bings of West Lothian, on which she is the acknowledged expert. Her report for West Lothian Council, which asked for guidance on what it should 'do' about the remaining oil-shale bings in its jurisdiction, persuaded them that for the sake not only of maintaining the highest biodiversity but also retaining the eloquence of those additions to the natural landscape, they should 'do' ... nothing. This they have now accepted. I wish that more local authorities could be persuaded of the benefits of *not* commoditizing and amenitizing the past.

This is not the first time I have been pleased and grateful to acknowledge the assistance and inspiration of Dr Bryan Lovell OBE. On this occasion, I have the pleasure of adding his wife Carol. Since I last acknowledged him, in my book *Supercontinent*, Bryan became not only President of the Geological Society, but also the unofficial high priest of the Hertfordshire Puddingstone. His thoughts on the geological story revealed by this unique deposit, and the lessons that it teaches not only society in general but the oil industry in particular, were first published in a three-part essay in *Geoscientist*. Those thoughts have since grown into a deservedly influential book, *Challenged by Carbon* (Cambridge University Press). Bryan has continued to be of incalculable help to me – and to anyone, indeed, who can help to get the message in the Puddingstone across to a wider audience. I feel I should not only acknowledge him personally, but, as it were, on behalf of our species: a grandiosity that will make him wince, but which I do not think is undeserved.

The repeated mention of *Geoscientist* may sound like advertising, but that is not my intention. It indicates how indebted I remain to it, the Society of whose Fellowship it is the independent monthly magazine and to its readers and contributors, for my continuing geological education.

When I explained what this book was to be about to my science-writer colleague Paul Sutherland (who occasionally goes by the sobriquet of *The Sun*'s Spaceman) he asked a perceptive and deliberately provocative question. In making this book a tribute to my mining, quarrying and stone-dressing forebears, he asked: 'Doesn't it make you feel guilty, the way our ancestors spent their entire lives grafting, and all we do is mess about with words?'

Well, yes, it does. But in a way, that is also the point. Had they not done what they did for their sons and daughters, I would not be able to do what I do – because that is the deal (or should be) between parents and children, between any generation and its successors. And that is, of course, the central message of this book. Times must change; but if each generation thinks not of itself but of its successors, holding the Earth in stewardship for them and not trashing it in a self-absorbed race for short-term gain in the here and now, our species stands a chance of stopping before it hits the wall it can already see coming. Only then, by ensuring that we pass on blessings and not curses to our offspring, will we ensure continuance. We must return, in a metaphorical way, to being ancestor worshippers.

My grateful thanks go to my editors at Granta, especially Bella Lacey and Christine Lo, whose perceptive guidance saved me from myself on innumerable occasions. Finally, I would like to thank Liz Hunt and all the team at Campion Gardens Nursing Home, Swansea, for their friendship, expert care, help and support for my father during his final year.

Ted Nield

Illustrations

Index